Lecture Notes in Engineering

The Springer-Verlag Lecture Notes provide rapid (approximately six months), refereed publication of topical items, longer than ordinary journal articles but shorter and less formal than most monographs and textbooks. They are published in an attractive yet economical format; authors or editors provide manuscripts typed to specifications, ready for photo-reproduction.

The Editorial Board

Lecture Notes in Engineering

Edited by C. A. Brebbia and S. A. Orszag

6

Myron B. Allen III

Collocation Techniques
for Modeling Compositional
Flows in Oil Reservoirs

Springer-Verlag Berlin Heidelberg GmbH 1984

Series Editors
C. A. Brebbia · S. A. Orszag

Consulting Editors
J. Argyris · K.-J. Bathe · A. S. Cakmak · J. Connor · R. McCrory
C. S. Desai · K.-P. Holz · F. A. Leckie · F. Pinder · A. R. S. Pont
J. H. Seinfeld · P. Silvester · P. Spanos · W. Wunderlich · S. Yip

Author
Myron B. Allen III
Department of Mathematics
University of Wyoming
Laramie, Wyoming
USA

ISBN 978-3-540-13096-3   ISBN 978-3-642-82213-1 (eBook)
DOI 10.1007/978-3-642-82213-1

2061/3020-543210

ABSTRACT

The numerical simulation of multiphase, multicomponent flows in porous media requires effective discrete techniques for solving the nonlinear partial differential equations governing transport of molecular species. Numerical models also demand reliable algorithms for computing the effects of interphase mass transfer on fluid properties. This investigation introduces a finite-element collocation method for solving the transport equations of compositional flows and advances a novel approach for improving phase-behavior calculations.

The collocation technique relies on a new method, called upstream collocation, to furnish convergent approximations to the equations of mass conservation. Through this method it is possible to construct collocation approximations analogous to the upwind finite-difference schemes dominating the literature on compositional modeling. A series of examples demonstrates the effectiveness of upstream collocation in related but more tractable flow problems.

The proposed thermodynamic algorithms use standard equation-of-state methods to compute geometric representations of the Maxwell sets of fluid mixtures. This approach replaces the solution of sensitive systems of nonlinear algebraic equations by a simple interpolation scheme during execution time. Since it is based on equation-of-state calculations, the interpolation scheme preserves thermodynamic consistency. Moreover, the new method mitigates the expense and convergence difficulties associated with the standard approach when more than one hydrocarbon phase is present.

FOREWORD

This investigation is an outgrowth of my doctoral dissertation at Princeton University. I am particularly grateful to Professors George F. Pinder and William G. Gray of Princeton for their advice during both my research and my writing.

I believe that finite-element collocation holds promise as a numerical scheme for modeling complicated flows in porous media. However, there seems to be a "conventional wisdom" maintaining that collocation is hopelessly beset by oscillations and is, in some way, fundamentally inappropriate for multiphase flows. I hope to dispel these objections, realizing that others will remain for further work.

The U. S. National Science Foundation funded much of this study through grant number NSF-CEE-8111240.

TABLE OF CONTENTS

CHAPTER ONE
THE PHYSICAL SYSTEM

1.1. Introduction.

Long before the oil price shocks of the past decade engineers recog-
nized a need for improved methods of exploiting petroleum resources.
Conventional production practices including waterflooding are so inefficient that, of all the oil discovered in the United States as of 1980,
less than 35 percent is identified as either having been recovered or
remaining as proved reserves (Doscher, 1980). By this count, about two
thirds of the nation's known original petroleum resources cannot be
produced using standard primary and secondary methods. Miscible gas
flooding, especially with CO_2 as the injected fluid, is one of the more
promising technologies for enhancing oil recovery and thus for shrinking
the gap between discovered resources and crude reserves (Holm, 1982).
Engineers designing miscible gas floods rely on mathematical models to
compare possible operating strategies and to estimate the amount and
timing of additional production. We shall examine new techniques applicable to the numerical simulation of miscible gas floods and similar
compositional flows in porous media.

How miscible gas flooding works.

The basic idea behind a miscible gas flood is to inject a relatively
cheap fluid, often CO_2 or propane-enriched natural gas, into a permeable, oil-bearing rock formation with the aim of driving the resident
oil toward producing wells. The most obvious principle by which
miscible gas flooding enhances recovery is a purely mechanical one:
injecting fluid into the porous reservoir causes an increase in pressure
drops between injection wells and production wells, resulting in greater
fluid velocities toward producers. This mechanical displacement is
common to all fluid injection schemes for enhanced oil recovery.

Miscible gas flooding offers additional mechanisms for improving oil production. In a well designed flood, the injected fluid mixes with the oil in place to form a zone in which the displacing fluid and the displaced fluid have very low interfacial tension. As this zone sweeps through the formation it moves oil that was previously trapped by the capillary forces present in the porous rock matrix. Because of its effects on capillarity, miscible gas flooding leaves less oil in the swept portions of the reservoir than recovery technologies, such as waterflooding, based on immiscible displacement.

The key to miscible displacement is the transfer of mass between the displacing and displaced fluid phases. As the fluids move with different velocities through the rock there is an exchange of molecular species in accordance with laws governing the compositions of coexisting phases. Thus, while the injected fluid initially may not be miscible with the reservoir oil, the interaction of the flow field and the fluid-phase thermodynamics leads to "developed" or "multiple-contact" miscibility. Holm (1976), Stalkup (1978), and Holm (1982) summarize the large body of literature describing this class of mechanisms.

An overview of mathematical modeling.

There are two essentially different approaches to modeling miscible gas floods. One of these is to forgo explicit simulation of interphase mass transfer, using as a surrogate any of several phenomenological mixing models coupled with a standard immiscible flow simulator (Lantz, 1970; Todd and Longstaff, 1972; Watkins, 1982). While this route is inexpensive and therefore quite popular, its success depends as much on felicitous choices of various fitting parameters as on the correct mathematical description of physical processes.

Here we shall be concerned with the second approach, namely, modeling both the flow field and the effects of interphase mass transfer.

Compositional reservoir simulators attempt to capture the complex inter-
actions between flow and thermodynamics in miscible gas floods, and
hence these models must cope with strongly nonlinear phenomena. The
first truly compositional simulators appeared in the American petroleum
engineering literature in the late 1960's and early 1970's (Price and
Donohue, 1967; Roebuck et al., 1969; Nolen, 1973; Van Quy et al., 1973).
Among the most recent and sophisticated of the reported compositional
simulators for miscible gas flooding are those of Kazemi et al. (1978),
Fussell and Fussell (1979), Coats (1980), Nghiem et al. (1981), and
Young and Stephenson (1982). Chapter Four of this investigation reviews
details of these models' structures. For now let us briefly note
current trends in the two major issues confronting designers of compositional simulators: the discretization of the differential equations
governing fluid motions and the numerical representation of fluid-phase
thermodynamics.

All of the simulators mentioned above use finite differences to
discretize both the space and time dimensions. While the use of finite
elements in petroleum reservoir simulation has shown steady progress
over the last decade, their application to compositional miscible gas
flood simulators has been sparse. It is particularly interesting that
finite-element collocation has received little attention in the petroleum industry compared to Galerkin finite-element techniques, despite
certain attractive features of the method. Chapter Three discusses
these matters more thoroughly.

Regarding the thermodynamic part of the problem, there appears to be
a trend toward the exclusive use of cubic equations of state to predict
fluid densities and compositions. All of the aforementioned simulators
reported since 1979 use equation-of-state methods. These methods have
the advantages of thermodynamic consistency and somewhat greater generality over the tabulated correlations used in earlier compositional
models. On the other hand, equation-of-state methods as commonly

implemented are expensive and exhibit poor convergence near critical
points of fluid mixtures, and these facts give cause for dissatisfaction
with the available techniques. Chapter Two treats these issues in some
detail.

Scope of the investigation.

The present study proposes new techniques for solving both the
thermodynamic constraints and the flow equations in the compositional
simulation of miscible gas floods. For the thermodynamic calculations
we shall construct a method for computing the compositions and saturations of coexisting fluid phases using an interpolation scheme in
conjunction with a cubic equation of state. This method, motivated by a
geometric view of equilibria in thermodynamic systems, furnishes a
simple and computationally reliable remedy to the expense and convergence difficulties associated with standard equation-of-state methods.
For the flow equations, we shall develop a new variant of finite-element
collocation that offers a viable alternative to finite-difference
methods. The collocation schemes presented in this study are novel in
that, unlike earlier methods, they give good results when applied to the
types of nonlinear, convection-dominated flows encountered in oil reservoirs.

This study has five chapters. The remainder of Chapter One establishes the physics governing miscible gas floods and notes some implications for the numerics that follow. Chapter Two reviews the thermodynamics of miscible gas floods in detail and presents the new
interpolation technique. Chapter Three introduces the collocation
method and demonstrates its applicability to equations governing flow
and transport in porous media. Chapter Four discusses the design of a
compositional simulator for miscible gas floods. Finally, Chapter Five
summarizes the results of the investigation and outlines possible directions for further work.

1.2. The reservoir and its contents.

This section briefly defines a petroleum reservoir in mathematical
terms suitable for use in the mechanical description that follows. The
definition is sufficiently broad to encompass fluid displacements
observed at the laboratory, pilot, and field scales, provided certain
underlying assumptions hold. The discussion that follows, including the
remainder of this chapter, is essentially a compilation of relevant
porous-media physics developed by many other researchers. There is no
essential any novelty in this review except perhaps in the attempt to
gather results and methods from a large and varied body of literature
into a consistent picture of miscible gas flood mechanics. While cited
references appear where the justification of facts or lines of reasoning
is not explicit, we shall not undertake an exhaustive literature review
on all of the topics discussed here.

Our point of view on porous-media physics is a macroscopic one. All
of the kinematic and dynamic quantities mentioned in this chapter therefore refer to a level of observation where a fluid-saturated porous
medium appears as a collection of superposed continua. Atkin and Craine
(1976) review the theory and history of this viewpoint, and their treatment and the work of Bowen (1980, 1982) guide much of the framework
outlined in this section and the next.

Let us assume that the reservoir consists of a body of homogeneous
porous matrix occupied by at most two fluids, a vapor and a liquid, to
which the matrix is permeable. Let us allow the compositions and densities of the fluids to vary in space and time. In practical terms these
assumptions exclude fractured porous media and such multi-fluid systems
as oil-gas-water reservoirs and liquid-liquid-vapor mixtures having more
than two fluid phases. Let us assume further that the reservoir is
isothermal, so that there is no need to account for material transfers
of energy through an energy balance equation.

In mathematical language, a reservoir is the closure Ω of a simply connected open subset of the Euclidean space $\mathbf{R}^d$, where $d = 1$, 2, or 3 depending on the dimensionality of the problem to be solved. Each neighborhood of any spatial point $\underset{\sim}{x} \in \Omega$ is occupied by matter from each element in a set Ξ of constituents. This set Ξ admits a decomposition $\Xi = \{1,2,\dots,N+1\} \times \{V,L,R\}$, where $\{1,2,\dots,N+1\}$ is the set of components or molecular species and $\{V,L,R\}$ is the set of phases, namely vapor, liquid, and rock. Thus a constituent is an ordered pair (i,α), identified as component i occurring in phase α.

Associated with each constituent (i,α) is a measure $\rho_{i\alpha}$, defined on the field of Lebesgue-measurable subsets of Ω and absolutely continuous with respect to Lebesgue measure. The measure $\rho_{i\alpha}$ is called the bulk molar density of species i in phase α, and its physical dimensions are [moles of (i,α) / volume]. In terms of $\rho_{i\alpha}$, the mixture density is

$$\rho_{mix} = \underset{i}{\Sigma} \underset{\alpha}{\Sigma} \rho_{i\alpha}$$

(1.2-1)

and the bulk density of phase α is

$$\rho_\alpha = \underset{i}{\Sigma} \rho_{i\alpha}$$

(1.2-2)

Also, associated with each phase $\alpha \in \{V,L,R\}$ is a continuous function $\phi_\alpha: \Omega \to [0,1]$, the volume fraction of α, having physical dimensions [volume of α / volume]. The set $\{\phi_V, \phi_L, \phi_R\}$ satisfies $\phi_V + \phi_L + \phi_R = 1$. Using the volume fractions and densities, it is possible to define several useful quantities as shown in Table 1-1.

TABLE 1-1

DEFINITIONS OF MASS-RELATED QUANTITIES

Symbol	Name	Definition	Physical Dimensions
ρ^α	Intrinsic molar density of phase, α	ρ_α/ϕ_α	[moles of α/volume of α]
ρ_i^α	Intrinsic molar density of constituent (i,α)	$\rho_{i\alpha}/\phi_\alpha$	[moles of (i,α)/volume of α]
ω_i^α	Mole fraction of species i in phase α	$\rho_i^\alpha/\rho^\alpha$	[moles of (i,α)/moles of α]
ω_i^{mix}	Mole fraction of species i in the mixture	$\rho_{mix}^{-1}\sum_\alpha \phi_\alpha \rho^\alpha \omega_i^\alpha$	[moles of i/moles of mixture]
ρ	Molar fluid density	$\phi(S_V\rho^V + S_L\rho^L)$	[moles of fluid/volume]
ω_i	Mole fraction of species i in the fluids	$\rho^{-1}\phi(S_V\rho^V\omega_i^V + S_L\rho^L\omega_i^L)$	[moles of i/moles of fluid]

For the rock-fluid system under consideration it is convenient to treat the rock phase separately, since its response to applied loads will be less important than the responses of the fluid phases. Call

$$\phi = \phi_V + \phi_L = 1 - \phi_R$$

(1.2-3)

the porosity, and define

$$S_V = \phi_V/\phi, \quad S_L = \phi_L/\phi$$

(1.2-4)

to be the saturations of vapor and liquid, respectively. Several useful quantities defined in terms of the saturations, ϕ, and previously defined variables also appear in Table 1-1. From the definitions and the fact that volume fractions sum to unity follow four restrictive equations:

$$\sum_i \omega_i = 1$$

(1.2-5a)

$$\sum_i \omega_i^V = 1$$

(1.2-5b)

$$\sum_i \omega_i^L = 1$$

(1.2-5c)

$$S_V + S_L = 1$$

(1.2-5d)

1.3. Reservoir mechanics.

Kinematics

At each time t there is associated with every spatial point $\underset{\sim}{x} \in \Omega$ a material point $\underset{\sim i}{X^\alpha}$ of each constituent $(i,\alpha) \in \Xi$. The mapping $\underset{\sim i}{x^\alpha}$: $(\underset{\sim i}{X^\alpha}, t) \mapsto x$ is the (i,α)-th motion, which we shall assume to be continuously differentiable with nonzero Jacobian determinant. Thus $\underset{\sim i}{x^\alpha}$ is at least locally invertible, with inverse $(\underset{\sim i}{x^\alpha})^{-1}$: $(\underset{\sim}{x}, t) \mapsto \underset{\sim i}{X^\alpha}$.

Given the motions of all of the constituents, it is possible to define various velocities that are useful in describing the behavior of the system:

$$\underset{\sim i}{v^\alpha}(\underset{\sim i}{X^\alpha}, t) = \partial_t \underset{\sim i}{x^\alpha}(\underset{\sim i}{X^\alpha}, t)$$

$$(1.3\text{-}1a)$$

is the velocity of constituent (i,α),

$$\underset{\sim}{v^\alpha} = \sum_i \rho_i^\alpha \underset{\sim i}{v^\alpha} / \rho^\alpha$$

$$(1.3\text{-}1b)$$

is the mean velocity of phase α,

$$\underset{\sim i}{u^\alpha} = \underset{\sim i}{v^\alpha} - \underset{\sim}{v^\alpha}$$

$$(1.3\text{-}1c)$$

is the diffusion velocity of species i in phase α,

$$\underset{\sim}{v}_{mix} = \sum_i \sum_\alpha \phi_\alpha \rho_i^\alpha \underset{\sim i}{v^\alpha} / \rho_{mix}$$

$$(1.3\text{-}1d)$$

is the barycentric velocity, and

$$\underset{\sim}{v}_i = \underset{\alpha \neq R}{\Sigma} \; \phi_\alpha \; \rho_i^\alpha \; \underset{\sim}{v}_i^\alpha \; / \; \rho$$

$$(1.3\text{-}1e)$$

is the mean velocity of species i in the fluids. From the definitions of ρ^α and $\underset{\sim}{v}^\alpha$ there follows

$$\underset{i}{\Sigma} \; \rho_i^\alpha \; \underset{\sim}{u}_i^\alpha = 0$$

$$(1.3\text{-}2)$$

Mass balance

Let us assume that the reservoir and its contents obey the mass balance. That is, for a given material volume $\Gamma \subset \Omega$,

$$d_t \; (\underset{i \; \alpha}{\Sigma \; \Sigma} \; \underset{\Gamma}{\int} \; \rho_i^\alpha \; dv) = 0$$

$$(1.3\text{-}3)$$

A standard argument (Eringen and Ingram, 1965) reduces this equation to

$$\underset{i \; \alpha}{\Sigma \; \Sigma} \; \underset{\Gamma \backslash \Sigma}{\int} \; [\partial_t \rho_i^\alpha + \nabla \bullet (\rho_i^\alpha \; \underset{\sim}{v}_i^\alpha)] \; dv$$

$$+ \underset{i \; \alpha}{\Sigma \; \Sigma} \; \underset{\Sigma}{\int} \; [\; \rho_i^\alpha \; (\underset{\sim}{v}_i^\alpha - \underset{\sim}{u}_\Sigma) \;] \bullet \; \underset{\sim}{n} \; ds = 0$$

$$(1.3\text{-}4)$$

In this equation, Σ is any surface in Γ, $\underset{\sim}{u}_\Sigma$ is the velocity of Σ, $\underset{\sim}{n}$ is a unit vector normal to Σ, and $[\bullet]$ signifies the jump in the quantity $(\bullet)$ across Σ. The surface integral will contribute to the left side of

(1.3-4) when Σ is a surface on which some of the densities ρ_i^α are discontinuous.

If (1.3-4) is valid for arbitrary material volumes Γ, no matter how small, then the following point balances hold:

$$\sum_i \sum_\alpha [\partial_t \rho_i^\alpha + \nabla \cdot (\rho_i^\alpha \underset{\sim}{v}_i^\alpha)] = 0 \quad \text{on } \Omega \setminus \Sigma$$

(1.3-5a)

$$\sum_i \sum_\alpha [\ \rho_i^\alpha (\underset{\sim}{v}_i^\alpha - \underset{\sim}{u}_\Sigma)\ ] \cdot \underset{\sim}{n} = 0 \quad \text{on } \Sigma.$$

(1.3-5b)

This system is equivalent to the following constituent balance equations

$$\partial_t \rho_i^\alpha + \nabla \cdot (\rho_i^\alpha\ \underset{\sim}{v}_i^\alpha) = \hat{\rho}_i^\alpha \quad \text{on } \Omega \setminus \Sigma$$

(1.3-6a)

$$[\ \rho_i^\alpha (\underset{\sim}{v}_i^\alpha - \underset{\sim}{u}_\Sigma)\ ] \cdot \underset{\sim}{n} = \hat{R}_i^\alpha \quad \text{on } \Sigma$$

(1.3-6b)

for all $(i,\alpha) \in \Xi$, provided we restrict the mass exchange terms $\hat{\rho}_i^\alpha$ and $\hat{R}_i^\alpha$ to obey

$$\sum_i \sum_\alpha \hat{\rho}_i^\alpha = \sum_i \sum_\alpha \hat{R}_i^\alpha = 0$$

(1.3-7)

Equations (1.3-6a) are partial differential equations governing the movement of matter in parts of the reservoir where the variables ρ_i^α

(and $\underset{\sim i}{v}{}^{\alpha}$) are smooth. Where jumps in densities occur, the conditions (1.3-6b) apply. From the mathematical viewpoint, we shall be concerned with solving (1.3-6a) numerically, but we shall admit solutions that satisfy these equations only in the weak sense. Hence the functions $\rho_i^{\alpha}(x,t)$ that we accept as solutions to the constituent balances may be discontinuous. However, we shall require such weak solutions to satisfy (1.3-6b) at their discontinuities.

Let us limit attention to a system in which the movements of matter obey a special set of simplifying assumptions. To begin with, the system undergoes no homogeneous chemical reactions; in other words, there is no exchange of mass among species within any phase. Thus,

$$\sum_{\alpha} \hat{\rho}_i^{\alpha} = 0, \qquad i = 1,\ldots,N+1$$

$$(1.3-8)$$

Moreover, the rock phase shares no species with the fluid phases, so that

$$\hat{\rho}_i^R = 0, \qquad i = 1,\ldots,N$$

$$(1.3-9)$$

$$\hat{\rho}_{N+1}^V = \hat{\rho}_{N+1}^L = 0$$

$$(1.3-10)$$

This assumption implies that $\hat{\rho}_i^V + \hat{\rho}_i^L = 0$. However, the fluids can exchange matter between themselves, so that in general

$$\sum_{i=1}^{N} \hat{\rho}_i^\alpha \neq 0, \quad \alpha = V, L$$

$$(1.3-11)$$

Thus the system to be modeled admits mass transfer between fluids but excludes adsorption, rock dissolution, and intraphase chemical reactions.

Since fluid-phase velocities in porous media are typically more accessible to measurement than constituent velocities, it is convenient to rewrite the constituent balance equations for fluid constituents in terms of v^α:

$$\partial_t(\phi \, S_\alpha \, \rho^\alpha \, \omega_i^\alpha) + \nabla\bullet(\phi \, S_\alpha \, \rho^\alpha \, \omega_i^\alpha \, \underset{\sim}{v}^\alpha) + \nabla\bullet\underset{\sim}{j}_i^\alpha = \hat{\rho}_i^\alpha,$$

$$i = 1, \ldots, N$$
$$\alpha = V, L$$

$$(1.3-12)$$

Here, $\underset{\sim}{j}_i^\alpha = \phi \, S_\alpha \, \rho^\alpha \, \omega_i^\alpha \, \underset{\sim}{u}_i^\alpha$ is the dispersion of species i with respect to the mean velocity of α. Summing (1.3-12) over the fluid phases and using the constraints on ω_i^α and $\hat{\rho}_i^\alpha$ gives the species balance equations,

$$\partial_t(\rho\omega_i) + \nabla\bullet[\phi(S_V \, \rho^V \, \omega_i^V \, \underset{\sim}{v}^V + S_L \, \rho^L \, \omega_i^L \, \underset{\sim}{v}^L)]$$

$$+ \nabla\bullet(\underset{\sim}{j}_i^V + \underset{\sim}{j}_i^L) = 0, \quad i = 1, \ldots, N$$

$$(1.3-13)$$

This leaves the rock balance equation,

$$\partial_t[(1 - \phi)\rho^R] + \nabla \cdot [(1 - \phi) \rho^R \underset{\sim}{v}^R] = 0$$

$$(1.3\text{-}14)$$

An analogous set of assumptions regarding the mass exchanges $\hat{R}{}_i^\alpha$ at included surfaces leads to the relationship $\hat{R}{}_i^V + \hat{R}{}_i^L = 0$. Thus, summing equations (1.3-6b) furnishes a jump condition corresponding to (1.3-13):

$$[\rho_i^V \underset{\sim}{v}_i^V + \rho_i^L \underset{\sim}{v}_i^L - (\rho_i^V + \rho_i^L)\underset{\sim}{u}_\Sigma] \cdot \underset{\sim}{n} = 0, \quad i = 1,\ldots,N$$

$$(1.3\text{-}15)$$

or

$$[\rho_i^V \underset{\sim}{v}_i^V + \rho_i^L \underset{\sim}{v}_i^L + (\underset{\sim}{j}_i^V/\phi S_V) + (\underset{\sim}{j}_i^L/\phi S_L) - (\rho_i^V + \rho_i^L)\underset{\sim}{u}_\Sigma] \cdot \underset{\sim}{n}$$

$$= 0, \quad i = 1,\ldots,N$$

$$(1.3\text{-}16)$$

Velocity field equations.

To avoid some of the complexities associated with modeling porous-media flows, let us assume that only the fluids move and that the density of the rock stays constant. Hence the rock is completely immobile, with $\underset{\sim}{v}^R = 0$ and both ρ^R and ϕ constant. Strictly speaking, this assumption is unrealistic, since in practical oilfield operations the pressure changes associated with pumping can cause detectable changes in porosity. These changes, however, often have small effects on fluid motions, and rock compressibilities for typical sandstones are frequently quite small ($\sim 10^{-10}$ Pa^{-1}, Collins, 1961, Chapter 1). The

assumptions that the rock is both chemically and mechanically inert
eliminate the need for solving the rock balance equation (1.3-14).

The fluid velocities pose greater difficulties. There now appears to
be no universally accepted, rigorous mechanical theory of multiphase
flows in porous media with interphase mass transfer. The only widely
used approach to this problem is to assume that the field equations for
simultaneously flowing fluid phases in porous media are extensions of
Darcy's law for single phase flow. Bowen (1980, 1982) develops this
theory for the case when no interphase mass transfer occurs.

The general law governing fluid-phase velocities in porous media is
the momentum balance, the local differential form of which is

$$\phi_\alpha \rho^\alpha (\partial_t \underset{\sim}{v}^\alpha + \underset{\sim}{v}^\alpha \bullet \nabla \underset{\sim}{v}^\alpha) = \nabla \bullet \underset{\approx}{t}^\alpha + \phi_\alpha \rho^\alpha \underset{\sim}{b}^\alpha + \overset{\wedge}{\underset{\sim}{p}}{}^\alpha$$

$$(1.3\text{-}17)$$

Here $\underset{\approx}{t}^\alpha$ is the stress tensor of phase α, $\underset{\sim}{b}^\alpha$ is the body force acting on
phase α, and $\overset{\wedge}{\underset{\sim}{p}}{}^\alpha$ is the net exchange of momentum into phase α from other
phases, subject to the restriction

$$\sum_\alpha \overset{\wedge}{\underset{\sim}{p}}{}^\alpha = \underset{\sim}{0}$$

$$(1.3\text{-}18)$$

Equation (1.3-17) holds for all phases, V, L, and R, even though only
the fluid phases are of interest here. A fairly simple set of assumptions reduces (1.3-17) to a multiphase version of Darcy's law for
isotropic media (Prevost, 1980; Hassanizadeh and Gray, 1980; Bowen,
1980, 1982). Without appealing to the generality of constitutive theory
(Ingram and Eringen, 1967), let us review these assumptions briefly.

First, assume that the fluids obey linear stress laws of the form

$$\underset{\approx}{t}^{\alpha} = - p_{\alpha} \underset{\approx}{I} + \sum_{\beta} \lambda^{\alpha\beta} \text{trace}(\underset{\approx}{d}^{\alpha}) \underset{\approx}{I}$$
$$+ \sum_{\beta} 2 \mu^{\alpha\beta} \underset{\approx}{d}^{\beta}, \qquad \alpha = V, L$$

$$(1.3\text{-}19)$$

where β ranges over all phases, p_{α} is the pressure of the fluid phase α, $\underset{\approx}{I}$ is the unit tensor, and $\underset{\approx}{d}^{\alpha} = \nabla \underset{\sim}{v}^{\alpha} + (\nabla \underset{\sim}{v}^{\alpha})^{\top}$. The coefficients $\lambda^{\alpha\alpha}$, $\mu^{\alpha\alpha}$ are intrinsic Lamé moduli for the fluid phase α, and coefficients of the form $\lambda^{\alpha\beta}$, $\mu^{\alpha\beta}$, $\alpha \neq \beta$, represent the effects of interphase tractions. Second, assume that the last two terms in (1.3-19) contribute negligibly to the fluid motions on the grounds that viscous effects are dominated by the effects of momentum exchanges with the rock matrix. This assumption reduces (1.3-19) to

$$\underset{\approx}{t}^{\alpha} = - p_{\alpha} \underset{\approx}{I}, \qquad \alpha = V, L$$

$$(1.3\text{-}20)$$

Third, assume that gravity is the only body force:

$$\underset{\sim}{b}^{\alpha} = g \nabla D, \qquad \alpha = V, L$$

$$(1.3\text{-}21)$$

where g is the magnitude of acceleration due to gravity (9.80 m/s), and D signifies the depth below some datum. Fourth, assume that the momentum exchanges take the form of isotropic Stokes drags:

$$\underset{\sim}{\hat{p}}^{\alpha} = \phi_{\alpha} \Lambda_{\alpha}^{-1} (\underset{\sim}{v}^{R} - \underset{\sim}{v}^{\alpha}) = - \phi_{\alpha} \Lambda_{\alpha}^{-1} \underset{\sim}{v}^{\alpha}, \qquad \alpha = V, L$$

$$(1.3\text{-}22)$$

Thus the effects of interfluid momentum exchanges are negligible compared with those of fluid-solid exchanges. Finally, assume that the inertial terms are negligible in the fluids, so that

$$\partial_t \underset{\sim}{v}^\alpha + \underset{\sim}{v}^\alpha \cdot \nabla \underset{\sim}{v}^\alpha = 0, \quad \alpha = V, L$$

(1.3-23)

Substituting assumptions (1.3-20) through (1.3-23) into the momentum balance (1.3-17) gives

$$0 = - \nabla p_\alpha + \rho^\alpha g \, \nabla D - \phi_\alpha \, \Lambda_\alpha^{-1} \, \underset{\sim}{v}^\alpha, \quad \alpha = V, L$$

(1.3-24)

The reciprocal of the Stokes drag coefficient is the mobility Λ_α of fluid phase α. The most common treatment of this parameter is to factor it as

$$\Lambda_\alpha = k_\alpha / \mu^\alpha$$

(1.3-25)

where k_α is the effective permeability of the rock matrix to phase α, having dimensions $[L^2]$, and μ^α is the dynamic viscosity of phase α. With this identification and the definition $S_\alpha = \phi_\alpha / \phi$, equation (1.3-22) becomes

$$\underset{\sim}{v}^\alpha = - k_\alpha (\phi \, S_\alpha \, \mu^\alpha)^{-1} \, (\nabla p_\alpha - \rho^\alpha g \, \nabla D)$$

(1.3-16)

which is the multiphase extension of Darcy's law for isotropic media.

1.4. Supplementary constraints.

Certain functional relationships hold among the variables describing the reservoir and its contents. These relationships supplement those given by the species balances, restrictive equations, and velocity field equations, providing information necessary to close the transport problem. Such relationships fall into two categories: thermodynamic constraints and constitutive laws.

Thermodynamic constraints.

The thermodynamic constraints govern the densities and compositions of the fluid phases as well as the relative amounts of the phases present at each point in the reservoir. In an isothermal system, these quantities depend on the overall composition of the fluid mixture at the given point and on the local pressures. Thus for example the molar density of a fluid phase α obeys a constraint of the form

$$\rho^\alpha = \rho^\alpha(\omega_i^\alpha, \ldots, \omega_{N-1}^\alpha, P_\alpha), \quad \alpha = V, L$$

$$(1.4\text{-}1)$$

Similarly, the fluid-phase molar compositions satisfy constraints of the form (Nikolaevskii and Somov, 1978)

$$\omega_i^\alpha = \omega_i^\alpha(\omega_1, \ldots, \omega_{N-1}, P_\alpha), \quad i = 1, \ldots, N-1, \quad \alpha = V, L.$$

$$(1.4\text{-}2)$$

In miscible gas floods relationships (1.4-1) and (1.4-2) for different fluid phases are not independent, since under appropriate conditions the densities and compositions of the fluids may become locally indistinguishable. When these conditions occur the interface between the phases vanishes and the fluids flow miscibly, allowing very efficient displacement of the liquid initially present. Indeed, the formation of a zone of such miscibly flowing fluids is the principal trait of a successful miscible gas flood. To model this phenomenon, the phase densities and compositions must satisfy

$$\lim_{cr}(\rho^V - \rho^L) = 0$$

(1.4-3a)

$$\lim_{cr}(\omega_i^V - \omega_i^L) = 0, \quad i = 1,\ldots,N-1$$

(1.4-3b)

Here "$\lim_{cr}$" signifies the limit as $(\omega_1,\ldots,\omega_{N-1},p_V)$ approaches a critical point, where phases become indistinguishable, from thermodynamic states inside the two-phase regime.

The relative amounts of phases present at any given point obey a constraint on the fluid saturations S_α. For computational purposes discussed in Chapter Two, however, this constraint is more conveniently expressed in terms of the mole fractions of the fluid mixture occuring as vapor or liquid. These quantities stand in direct correspondence to the saturations through the definition

$$Y_\alpha = \rho^\alpha S_\alpha/(\rho^V S_V + \rho^L S_L), \qquad \alpha = V,L.$$

(1.4-4)

The phase mole fractions satisfy the restrictive equation

$$Y_V + Y_L = 1$$

(1.4-5)

and obey a thermodynamic constraint of the form

$$Y_V = Y_V(\omega_1, \ldots, \omega_{N-1}, p_V)$$

(1.4-6)

The specific forms of the thermodynamic constraints (1.4-1), (1.4-2), and (1.4-6) depend on the choice of thermodynamic data. We shall use constraints derived from standard equation-of-state methods, which Chapter Two discusses in more detail. For the present it suffices to note that in the usual equation-of-state approach the thermodynamic constraints assume the mathematical form of a collection of nonlinear algebraic equations that are equivalent to the explicit forms stated above but are implicit in the densities, fluid-phase compositions, and phase mole fractions.

Constitutive laws.

The constitutive laws govern the behavior of certain parameters of the transport problem which may vary as the flow field evolves. These parameters are dispersion, capillary pressure, effective permeabilities, and fluid viscosities.

Dispersion.

Dispersion is arguably the most poorly understood macroscopic phenomenon in porous media physics, and and there exists a correspondingly large literature on the subject (see Perkins and Johnson, 1963; Greenkorn and Kessler, 1969; Nunge and Gill, 1969, Fried and Combarnous, 1971, and Fried, 1975, Chapter 2, for reviews). Recent studies have shown some emphasis on relating the macrocsopic features of dispersion to the microscopic structure of fluid-saturated porous media (see, for examples, Carbonell and Whitaker, 1982; Mohanty and Salter, 1982; Smith and Schwartz, 1980). There is fair agreement that for flows of a single fluid phase at a fixed scale dispersion obeys a phenomenological constitutive law of Fick's type (Van Quy et. al, 1972),

$$\underset{\sim}{j}_i = - \phi \, \rho \, M \, \underset{\approx}{D}_i \cdot \nabla(M_i \omega_i / M)$$

$$(1.4\text{-}7)$$

where M_i is the molar mass of species i and $M = (\Sigma_{j=1}^{N} \omega_j / M_j)^{-1}$ is the molar mass of fluid. However, no simple choice of functional dependence for the dispersion tensor $\underset{\approx}{D}_i$ has escaped criticism on both experimental and theoretical grounds. Moreover, there has been no empirically tested extension of (1.4-7) to multiphase flows. Since the dispersion tensor must account for such disparate microscopic phenomena as molecular diffusion, Taylor diffusion (Taylor, 1953), stream splitting, and tortuosity of the matrix, it is perhaps no wonder that the most appropriate functional form remains a mystery.

The most common treatment of dispersion in compositional models of miscible gas floods to date has been to ignore it. Of the major compositional simulators reported in the American petroleum engineering literature, only that of Van Quy et al. (1972) reports the use of dispersion coefficients. Although data quantifying dispersion in multiphase, multicomponent flows are scarce, some experiments suggest that this class of phenomena can have measurable effects in laboratory- and field-scale floods (Watkins, 1978; Yellig and Baker, 1981). The most straightforward extension of the single-phase law (1.4-7) is

$$\underset{\sim}{j}_i^{\alpha} = - \phi \, S_{\alpha} \, \rho^{\alpha} \, M^{\alpha} \, \underset{\approx}{D}_i^{\alpha} \cdot \nabla(M_i \omega_i^{\alpha} / M^{\alpha})$$

$$(1.4\text{-}8)$$

However attractive equation (1.4-8) may be on theoretical grounds, it remains untested. It is worth noting that in many applications the effects of dispersion may be quite small (Collins, 1961, Chapter 8), and in fact the fully compositional model developed in Chapter Four neglects the phenomenon entirely.

Capillary pressure.

The capillary pressure relates the macroscopic pressure in the liquid to that in the vapor:

$$P_{CVL} = P_V - P_L$$

$$(1.4-9)$$

This difference owes its existence to the interfacial tension between the fluid phases, a molecular phenomenon (Gubbins and Haile, 1977; Davis and Scriven, 1980), and to the microscopic geometry of the fluid interface in the interstices of the porous matrix (Morrow, 1969). Thus the actual physics of capillarity are quite complicated, and its details are not wholly accessible to the macroscopic level of observation to which the equations of this chapter pertain.

In bench- or field-scale studies it is most practical to measure capillary pressures as functions of macroscopic flow parameters, using the experimental values to define empirical capillary pressure functions applicable in the velocity field equations. There is general agreement that for two fluids of fixed compositions the capillary pressure depends on the local values and history of the saturations (Morrow, 1969). For flows in which saturations change monotonically, as in strict imbibition or strict drainage, the capillary pressure of fixed-composition fluids is a unique function of saturation for each initial state of the porous medium. In miscible gas floods, however, the fluid compositions vary in time and space. To account for these changes, it is convenient to quantify the interfacial tension of the fluid mixture as a function of thermodynamic variables:

$$\sigma = \sigma(\rho^V, \omega_1^V, \ldots, \omega_{N-1}^V, \rho^L, \omega_1^L, \ldots, \omega_{N-1}^L)$$

(1.4-10)

This quantity has the dimensions [energy/area], or $[M/T^2]$. The capillary pressure then has the functional form

$$P_{CVL} = P_{CVL}(S_V, \sigma)$$

(1.4-11)

As $\sigma \to 0$ the interface between vapor and liquid disappears and the displacement occurs miscibly; this happens at critical points. From this consideration it is clear that

$$\lim_{\sigma \to 0} P_{CVL} = 0$$

(1.4-12)

a condition that parallels equations (1.4-3). In the absence of experimental data we shall compute interfacial tensions using the Sugden-Macleod correlation (Reid et al., 1977, Chapter 12).

Effective permeabilities.

Like capillary pressure, effective permeabilities are also macroscopic manifestations of the effects of interstitial geometry and interfacial tension on fluid flows in porous media, and the caveats regarding

measurements at the bench or field scale hold here as well. It is
customary to factor the effective permeability of each fluid phase α as
$k_{\alpha} = kk_{r\alpha}$, where k is the absolute permeability, a characteristic of the
rock, and $k_{r\alpha}$ is the relative permeability, a phenomenological factor
accounting for the influence of the other fluid phase on the flow of α
and obeying $0 \leq k_{r\alpha} \leq 1$. I shall assume that k is a uniform, constant
scalar, thereby disregarding the possible effects of anisotropy or
pressure-induced matrix deformations on the permeability as well as the
Klinkenberg effect, an enhancement of permeability to vapor attributed
to slipping at the walls of the matrix (Collins, 1961, Chapter 3).

For relative permeabilities the appropriate constitutive laws are not
so simple. The influence of one fluid phase, say β, on the other, say
α, in general comprises a complicated set of microscopic phenomena.
Under the assumption that fluid-fluid tractions are small, we may
consider the main effect to be the obstruction of channels to the flow
of α owing to their occupation by β in a configuration depending on the
interfacial tension. While relative permeabilities have been thoroughly
studied for immiscible displacements without interphase mass transfer
(see Scheidegger, 1974, Chapter 10 for a review), there is little experimental information on relative permeabilities applicable to miscible
gas floods. Noteworthy exceptions are the studies of Bardon and
Longeron (1980), who examined gas-oil systems, and Amaefule and Handy
(1982), who used aqueous solutions of surfactants to displace refined
hydrocarbon mixtures. These investigations show that, for systems in
which interphase mass transfer changes the compositions of the fluids,
the relative permeabilities for monotonic displacements depend on
saturations and interfacial tensions:

$$k_{r\alpha} = k_{r\alpha}(S_V, \sigma), \quad \alpha = L, V.$$

$$(1.4\text{-}13)$$

To be consistent with the theory of single-phase flow in porous media, these functions must satisfy $0 \leq k_{rV} + k_{rL} \leq 1$. Moreover, as $\sigma \to 0$ the vapor-liquid interface disappears, and the single-phase version of Darcy's law applies:

$$\underset{\sim}{v} = - (k/\phi\mu)(\nabla p - \rho g \nabla D)$$

$$(1.4-14)$$

Therefore, as $\sigma \to 0$, $k_{rV} + k_{rL} \to 1$ for any saturation. Bardon and Longeron report that as σ becomes very small the relative permeability curves approach straight lines with positive and negative unit slope on the saturation interval $[0,1]$. This behavior occurs near critical points in miscible gas floods.

Viscosities.

Let us assume that viscosities are functions of pressure and phase composition:

$$\mu^\alpha = \mu^\alpha (p_\alpha, \omega_1^\alpha, \ldots, \omega_{N-1}^\alpha)$$

$$(1.4-15)$$

In the absence of experimental data a correlation developed by Lohrenz, Bray, and Clark (1964) gives fairly accurate predictions of mixture viscosities.

1.5. Governing equations.

The balance laws together with the restrictive equations and supplementary constraints combine to form a system of equations that must be solved to predict the performance of miscible gas floods. Since this system is nonlinear and usually quite complicated, numerical approximation offers the only hope for producing solutions in a practical fashion. To motivate the choices of numerical procedures in the following chapters, let us close this chapter with a formal assembly of the equation set to be solved and a discussion of some of its mathematical aspects.

Form of the system.

The basic transport equation governing the distribution of any fluid-phase component i in the reservoir is the mass balance augmented by the velocity field equations and constitutive laws. Thus, substituting (1.3-26) and (1.4-8) into (1.3-13) gives

$$\partial_t (\rho \omega_i) - \nabla \bullet [\Lambda_V \, \rho^V \, \omega_i^V \, (\nabla p_V - \rho^V g \, \nabla D)$$

$$+ \Lambda_L \, \rho^L \, \omega_i^L \, (\nabla p_L - \rho^L g \, \nabla D) - \nabla \bullet (\underset{\sim}{j_i^V} + \underset{\sim}{j_i^L}) = 0$$

$$i = 1, \ldots, N$$

$$(1.5-1)$$

Using the capillary pressure $p_{CVL} = p_V - p_L$ and calling $\gamma_i' = (\Lambda_V \, \omega_i^V \, \rho^V + \Lambda_L \, \omega_i^L \, \rho^L)g$ reduces these equations to

$$\partial_t(\rho\omega_i) - \nabla \cdot [(\Lambda_V \, \rho^V \, \omega_i^V + \Lambda_L \, \rho^L \, \omega_i^L) \, \nabla p_V + \gamma_i{}' \, \nabla D$$

$$- \Lambda_L \, \rho^L \, \omega_i^L \, \nabla p_{CVL}] - \nabla \cdot (\underset{\sim}{j}_i^V + \underset{\sim}{j}_i^L) = 0,$$

$$i = 1, \ldots, N$$

$$(1.5\text{-}2)$$

Henceforth let us assume that the flow is one-dimensional, that is, that all variables are uniform along two Cartesian axes and vary only along x. Integrating (1.5-2) across the directions of uniformity and denoting the cross-sectional area by $A(x)$ then yields

$$A \, \partial_t(\rho\omega_i) - \partial_x[(T_V \, \omega_i^V + T_L \, \omega_i^L) \, \partial_x p_V + \gamma_i \, \partial_x D$$

$$- T_L \, \omega_i^L \, \partial_x p_{CVL}] - \partial_x[A(j_i^V + j_i^L)] = 0,$$

$$i = 1, \ldots, N$$

$$(1.5\text{-}3)$$

where

$$T_\alpha = A \, \Lambda_\alpha \, \rho^\alpha, \quad \alpha = V,L$$

$$(1.5\text{-}4a)$$

is the transmissibility of phase α and

$$\gamma_i = A \, \gamma_i{}'$$

$$(1.5\text{-}4b)$$

In addition to the N equations (1.5-3) we have the four restrictive equations

$$\sum_{i=1}^{N} \omega_i^V = \sum_{i=1}^{N} \omega_i^L = \sum_{i=1}^{N} \omega_i = S_V + S_L = 1$$

$$(1.5-5)$$

the 2N + 1 thermodynamic constraints

$$\rho^\alpha = \rho^\alpha (\omega_1^\alpha, \ldots, \omega_{N-1}^\alpha, P_\alpha), \quad \alpha = V, L$$

$$(1.4-1)$$

$$\omega_i^\alpha = \omega_i^\alpha (\omega_1, \ldots, \omega_{N-1}, P_\alpha), \quad i = 1, \ldots, N-1, \ \alpha = V, L$$

$$(1.4-2)$$

$$S_V = S_V(\omega_1, \ldots, \omega_{N-1}, P_V)$$

$$(1.5-6)$$

(the last being equivalent to (1.4-6) through the definition (1.4-4)), the definitions of ρ and g, and constitutive laws sufficient to determine T_V, T_L, P_{CVL}, j_i^V and j_i^L, $i = 1, \ldots, N$. Given the problem geometry (A(x) and D(x)) and appropriate boundary and initial data, equations (1.5-3), (1.5-5), (1.4-1), (1.4-2), and (1.5-6) constitute a set of 3N + 5 equations in the 3N + 5 unknowns $\{\omega_1, \ldots, \omega_N, \ \omega_1^V, \ldots, \omega_N^V, \ \omega_1^L, \ldots, \omega_N^L,$ $P_V, S_V, S_L, \rho^V, \rho^L\}$. Finally, to accommodate the event that spatially discontinuous solutions may arise, we have the jump conditions (1.3-16).

Weak solutions.

The possibility of discontinuous solutions raises the issue of weak solutions to the governing partial differential equations (1.5-3), since functions satisfying these equations in the literal sense cannot be discontinuous. For a function $\omega(x,t)$ to be a weak solution of a conservation law $\partial_t f_1(\omega) + \partial_x f_2(\omega) = 0$ on an (x,t)-domain $\Omega \times \Theta$, the following integral equation must hold for any function $g \in C^\infty(\Omega \times \Theta)$ having compact support in $\Omega \times \Theta$ (Chorin and Marsden, 1979, Chapter 3; Birkhoff, 1983):

$$\int_{\Omega \times \Theta} [f_1(\omega) \, \partial_t g + f_2(\omega) \, \partial_x g] \, dxdt = 0$$

$$(1.5-7)$$

This criterion admits discontinuous functions $\omega(x,t)$ but reduces to the original partial differential equation when ω is sufficiently smooth to satisfy the latter.

Equations (1.5-3) are fairly complex, and it is not at all clear on inspection whether one can reasonably expect discontinuous solutions to arise. There are, however, several simplified versions of these equations which have been shown to exhibit discontinuous solutions. Let us review two such simplifications: the Buckley-Leverett saturation equation and a generalization of the Buckley-Leverett theory due to Helfferich (1981, 1982).

The Buckley-Leverett problem

The Buckley-Leverett saturation equation (Buckley and Leverett, 1942) models the incompressible flow of two immiscible fluid phases in a homogeneous porous medium. The equation arises from a set of species balances of the form (1.5-3) under the further assumptions that $N = 2$ and neither species is shared between phases. Letting $\omega_1^L = \omega_2^V = 0$, then, we have from (1.5-3)

$$A \, \partial_t(\phi \, \rho^V \, S_V) - \partial_x(T_V \, \partial_x p_V + T_V \, \rho^V \, g \, \partial_x D) = 0$$

$$(1.5-8a)$$

$$A \, \partial_t (\phi \, \rho^L \, S_L) - \partial_x [T_L (\partial_x P_V - \partial_x P_{CVL}) + T_L \, \rho^L \, g \, \partial_x D] = 0$$

$$(1.5\text{-}8b)$$

Assuming also that gravity and capillarity have negligible effects, that variations in ρ^V and ρ^L are negligible, and that A is constant and uniform allows us to rewrite (1.5-8) as

$$\phi \, \partial_t S_V + \partial_x q_V = 0$$

$$(1.5\text{-}9a)$$

$$\phi \, \partial_t (1 - S_V) + \partial_x q_L = 0$$

$$(1.5\text{-}9b)$$

where $q_\alpha = - \Lambda_\alpha \partial_x P_V$ is the flow rate of phase α, $\alpha = V, L$. In the case where the total flow rate $q = q_V + q_L$ is constant we need only solve one of these equations, say the first. Since $q_V = \Lambda_V q / (\Lambda_V + \Lambda_L)$ this reduces to the hyperbolic conservation law

$$\partial_t S_V + \partial_x (q \, \phi^{-1} f_V) = 0$$

$$(1.5\text{-}10)$$

where the fractional flow function

$$f_V = \Lambda_V / (\Lambda_V + \Lambda_L) = f_V(S_V)$$

$$(1.5\text{-}11)$$

denotes the volume fraction of the flowing stream occupied by vapor.

The nature of the fractional flow function is, of course, crucial to the behavior of solutions to equation (1.5-10). Although the precise form of $f_V(S_V)$ depends on the particular rock and fluids studied, there

are qualitative features common to most fractional flow functions for
immiscible displacements, as drawn in Figure 1-1. First, f_V vanishes
for vapor saturations less than an irreducible vapor saturation S_{VR} and
equals unity for $S_V \geq 1 - S_{LR}$, S_{LR} being the residual liquid saturation.
These "endpoints" S_{VR} and $1 - S_{LR}$ are constants characteristic of the
given vapor-liquid-rock mixture when phase compositions do not vary.
Second, f_V is often not a convex function over its support $[S_{VR}, 1]$:
typically it is S-shaped, with an inflection point where its slope has a
maximum. Finally, $f_V'(S_V)$ exists throughout $[0,1]$ and vanishes at the
endpoints S_{VR} and $1 - S_{LR}$.

These peculiarities of $f_V(S_V)$ can lead to discontinuous solutions
$S_V(x,t)$ for Cauchy problems with initial data of the form

$$S_V(0,t) = S_1, \qquad t \geq 0$$

$$S_V(x,0) = S_2(x), \quad x > 0$$

$$(1.5\text{-}12)$$

defined on the (x,t)-domain $\Omega \times \Theta = [0,\infty) \times [0,\infty)$. To see this, observe
that (1.5-10) has the characteristic equation

$$(dx/dt)|_{S_V} = q\phi^{-1}(df_V/dS_V)$$

$$(1.5\text{-}13)$$

which governs loci of constant S_V. Since f_V is not not convex, df_V/dS_V
is not monotonic, and so there are speeds at which several distinct
saturations may propagate. Thus it is possible, for example, for the
locus of some large value of S_V to overtake that of a small value
initially ahead of it. Depending on the initial data prescribed, then,
literal application of (1.5-13) may eventually lead to multiple values

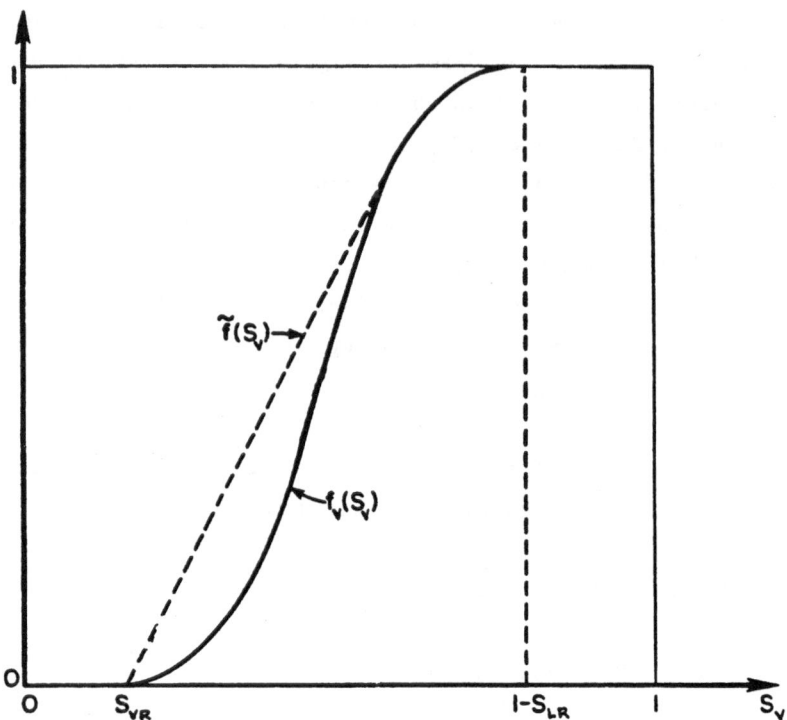

Fig. 1-1:   A typical fractional flow function $f_V(S_V)$ and
its convex hull $\tilde{f}(S_V)$.

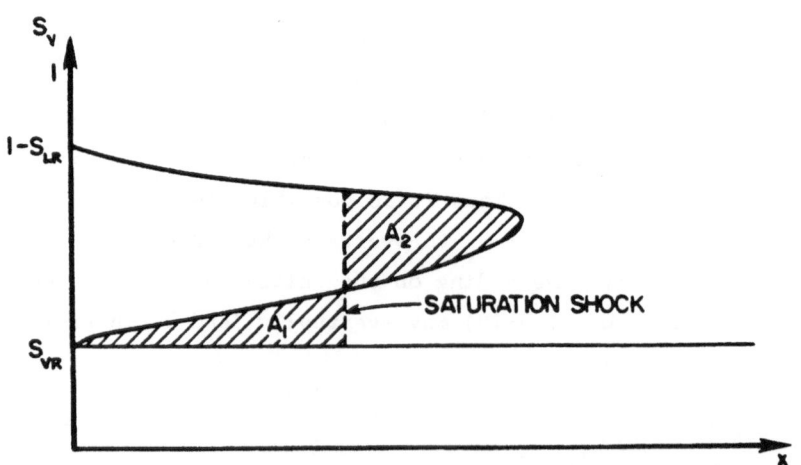

Fig. 1-2:   Triple-valued saturation predicted by the Buckley-
Leverett characteristic equation.

of vapor saturation at certain points in Ω, as shown in Figure 1-2. One
way to resolve this apparent paradox, as Welge (1952) shows, is to
replace f_V by its convex hull, drawn as the dashed line in Figure 1-1.
Welge's construction yields the single-valued but discontinuous saturation profile drawn in Figure 1-2. At smooth parts of the solution S_V,
any locus of constant saturation advances with the speed given by
(1.5-13), while discontinuities move with speeds satisfying the jump
condition (1.3-16). For the vapor-phase component this reduces to

$$[\; S_V \; v^V - S_V \; u_\Sigma \;] = 0$$

$$(1.5\text{-}14)$$

or, since $v^V = q f_V / \phi S_V$,

$$u_\Sigma = q \; \phi^{-1} \; [f_V]/[S_V]$$

$$(1.5\text{-}15)$$

The Buckley-Leverett-Welge solution is a weak solution but not a
classical solution, since it is discontinuous. Nonetheless, it is the
physically correct solution to the Buckley-Leverett problem.

Helfferich's theory.

Recently, Helfferich (1981, 1982) has presented a unified generalization of the Buckley-Leverett theory to multicomponent flows in porous
media with interphase mass transfer, and this work sheds some light on
the equations governing miscible gas floods. Helfferich bases his
approach on simplified versions of the species balances; we shall derive
an equivalent set of equations from (1.5-3).

Assume, as in the Buckley-Leverett problem, that gravity and capillarity have negligible effects, that A does not vary, and further that dispersion is negligible for all species. Under these hypotheses (1.5-3) becomes

$$\partial_t(\rho\omega_i) + \partial_x[q(f_V\omega_i^V + f_L\omega_i^L)] = 0, \quad i = 1,\ldots,N$$

$$(1.5\text{-}16)$$

where now $q_\alpha = - \rho^\alpha\Lambda_\alpha \, \partial_x P_V$ is the flow rate of phase α, $\alpha = V,R$, $q = q_V + q_L$ is the total flow rate, and $f_\alpha = q_\alpha/q$ is the fractional flow of phase α, $\alpha = V,L$. Calling $f_i = f_V\,\omega_i^V + f_L\,\omega_i^L$ and assuming that variations in molar fluid density in space and time are negligible, we can rewrite (1.5-16) as

$$\partial_t\omega_i + \partial_x(q \, \phi^{-1} f_i) = 0, \quad i = 1,\ldots,N$$

$$(1.5\text{-}17)$$

Now from the identity $f_V + f_L = 1$ it follows that $\Sigma_{i=1}^N f_i = 1$, and consequently only $N - 1$ of the equations (1.5-17) are independent.

For each species i, equation (1.5-17) is identical in form to the Buckley-Leverett saturation equation (1.5-10). By inspection, then, loci of constant ω_i move at the speed

$$(dx/dt)|_{\omega_i} = q\rho^{-1}(df_i/d\omega_i), \quad i = 1,\ldots,N\text{-}1$$

$$(1.5\text{-}18)$$

and discontinuities in ω_i travel at the speed

$$u_{\Sigma_i} = q\rho^{-1} [f_i]/[\omega_i], \quad i = 1,\ldots,N\text{-}1$$

$$(1.5\text{-}19)$$

By definition, a given composition $(\omega_1, \ldots, \omega_{N-1})$ propagates as a coherent wave provided all of the variables ω_i at a given time and place (x,t) in the reservoir move with the same speed; in symbols,

$$q\rho^{-1}(df_i/d\omega_i) = \lambda, \quad i = 1, \ldots, N-1$$

$$(1.5\text{-}20a)$$

Similarly, for a discontinuity to be coherent, we must have

$$q\rho^{-1} [\, f_i \,]/[\, \omega_i \,] = \Lambda, \quad i = 1, \ldots, N-1$$

$$(1.5\text{-}20b)$$

Consider the case when $N = 3$. Equations (1.5-20) become $q\,\rho^{-1}df_i = \lambda\,d\omega_i$ for $i = 1,2$, which by the chain rule can be written as

$$[\, \lambda\,I - q\rho^{-1}\,A\,]\,d\underset{\sim}{\omega} = 0$$

$$(1.5\text{-}21)$$

where A is the matrix whose elements are $\partial f_i/\partial\omega_j$, $i,j = 1,2$, I is the identity matrix, and $d\underset{\sim}{\omega} = (d\omega_1, d\omega_2)^T$. Equation (1.5-21) has nontrivial solutions only when the characteristic equation $\det(\lambda\,I - q\,\rho^{-1}\,A) = 0$ holds, that is, when λ is a root to the quadratic equation

$$\lambda^2 - \lambda\,q\,\rho^{-1}\,\text{trace}\,A + q^2\rho^{-2}\,\det\,A = 0$$

$$(1.5\text{-}22)$$

In this case, then, there will typically be two characteristic speeds of coherence λ_{max}, λ_{min} for a given composition (ω_1, ω_2) at a given point (x,t). The corresponding solutions of (1.5-21) define the tangents $(d\omega_1/d\omega_2)_{max}$, $(d\omega_1/d\omega_2)_{min}$ to the characteristic curves in composition space along which the λ_{max}- and λ_{min}-waves travel, respectively.

In actual problems the fractional flows f_i, and hence the matrix $\mathbf{A}$, are complicated functions of composition, even under the simplifying assumptions used to derive equations (1.5-17). We must therefore allow that the qualitative structures of the solutions to this problem may be quite different for various choices of thermodynamic and initial data. Helfferich (1982), however, adduces simple examples showing that discontinuous solutions can arise for physically reasonable problems in which $N = 2$ or 3, and his reasoning extends to larger numbers of species.

Implications for numerical solution.

The analyses of the Buckley-Leverett problem and Helfferich's theory have implications regarding the choice of numerical approximations to the full system (1.5-3). In particular the behaviors of the simplified systems suggest that the full system may possess solutions having steep gradients or discontinuities in composition. Indeed, finite-difference studies of compositional reservoir flows have borne out this expectation (see, for example, solutions plotted in Van Quy et al., 1972; Coats, 1980; Nghiem et al., 1981). The possibility of such shock-like solutions demands special numerical treatment of the governing equations to ensure physically correct approximations.

The complexity of the full system of species balances hinders rigorous analysis of its solutions. There do not even exist published proofs of existence or uniqueness of solutions to the full compositional equations, although Isaacson (1981) and Temple (1981) have produced a global existence proof for Cauchy problems on a simplified analog of (1.5-3) given initial data of bounded variation. Thus, while solving the compositional equations numerically is a problem of considerable practical importance, it is also a task lacking somewhat in mathematically rigorous support. We must therefore limit our discussion of the numerical solution to (1.5-3) to heuristic remarks based on analogies with the simpler cases.

The formation and persistence of steep composition gradients occur in problems where the dissipative influences of species dispersion and capillary pressure gradients are dominated by convection owing to applied pressure gradients. Such conditions are common in oilfield practice. When dispersion and capillarity are absent the governing equations take the form of hyperbolic conservation laws, and, as the simplified analyses show, we should expect discontinuities to form. If the dissipative terms are nonzero but very small, then because of their functional dependences on composition gradients their influences will be detectable only in the very near vicinity of steep portions of the solution. A discrete approximation to such terms may miss these influences altogether when their spatial extent is significantly smaller than the mesh of the spatial grid, and poor numerical approximations may result. In fact it is possible for consistent, apparently stable numerical schemes to fail to converge when applied to the Buckley-Leverett problem (Mercer and Faust, 1977; Allen and Pinder, 1982) or to the pressure-saturation equations governing two-phase immiscible displacements (Aziz and Settari, 1979, Chapter 5).

These convergence difficulties are symptoms of incompletely posed problems. The weak form (1.5-7) of a hyperbolic conservation law together with its jump condition are not sufficient to determine unique discontinuous solutions to otherwise well-posed Cauchy problems (Chorin and Marsden, 1979, Chapter 3). There is an additional constraint needed to close such problems, and it may be stated in several ways. Among them are the following:

(i)   The solution must depend continuously on the initial data, so that characteristics on both sides of any discontinuity must intersect the initial curve.

(ii) The solution must satisfy Oleinik's "condition E" (Oleinik, 1963b), a geometric constraint for nonconvex flux functions that reduces to the Welge construction in the Buckley-Leverett problem.

Another equivalent of this additional constraint which is useful for discrete approximations is the "vanishing viscosity" condition (Oleinik, 1963a; Lax, 1957 and 1972):

(iii) The solution must be the limit of solutions, for the same data, to a parabolic equation differing from the hyperbolic one by a dissipative second-order term (capillarity or dispersion, in our case) of vanishing influence.

All of the compositional simulators cited above impose the vanishing viscosity condition numerically through the use of upstream-weighted difference approximations to the flux term. Chapter Three describes a finite-element collocation scheme for (1.5-3) that imposes the vanishing viscosity condition in an analogous fashion.

CHAPTER TWO
REPRESENTING FLUID-PHASE BEHAVIOR

The key to the attractiveness of miscible gas flooding compared with immiscible flood technologies is interphase mass transfer. Under the right conditions of pressure, temperature, and composition, the exchange of species across the phase boundary in a miscible gas flood leads to the formation of a zone in which the displacing fluid and the displaced fluid are very similar. As described in Section 1.1, the rock swept by such a zone has very low residual oil saturation, implying more efficient oil recovery overall.

The transfer of mass among fluid phases in a miscible gas flood is a complicated set of kinetic phenomena driven by intermolecular forces and macroscopic transport phenomena. Mathematical models of such compositional flows typically invoke an assumption of "local thermodynamic equilibrium" to warrant the use of equilibrium methods in computing fluid-phase densities and compositions. In this approach one neglects the kinetics of interphase mass transfer, instead imposing thermostatic constraints at each location and instant in the flow field. The approach is computationally convenient, especially when one uses an equation of state to predict fluid-phase properties. This chapter briefly discusses the physical significance of the equilibrium approach through a thermodynamic framework that is consistent with established thermostatic results. We shall also review the calculations needed to apply these results with an equation of state, noting some of the undesirable features of the numerics. Finally, we shall examine a simple variant of the standard equation-of-state approach that retains most of its advantages but avoids its most salient computational shortcomings.

2.1. Thermodynamics of the fluid system.

It is clear that no transient system will obey the assumption of
thermostatic equilibrium in any strict sense. However, the assumption
has empirical support (see, for example, Raimondi and Torcaso, 1965) in
the sense that its predictions agree well with the results of flow
experiments. It is possible, moreover, to reconcile the notions of
equilibrium thermostatics with the description of many transient flows
using a dynamic interpretation of Gibbs' theory developed by Gilmore
(1981). Appendix B reviews the application of this interpretation to
the thermodynamics of miscible gas floods. This section summarizes in
somewhat less technical language the link between the geometry of
equilibria and the algebraic descriptions that we owe to Gibbs (1876 and
1878).

Notions of equilibrium.

The principal thermodynamic variables in an isothermal multicomponent
flow belong to two sets. The first, which we shall identify as the set
of control variables, is $\{\omega_1, \ldots, \omega_{N-1}, V\}$, where ω_i is the mole fraction
of species i in the fluid mixture and V is the molar fluid volume, that
is, the reciprocal of molar fluid density ρ. The second set of
variables, which we shall designate the state variables, is
$\{\eta_1, \ldots, \eta_{N-1}, -p\}$, where η_i is the modified chemical potential of
species i and p is the pressure. The variables η_i are different from
the customary chemical potentials μ_i, the two sets being related at
equilibrium by the equation $\eta_i = \mu_i - \mu_N$. While Appendix B derives
results in terms of the η_i for technical reasons, one can readily translate the derived relationships to equations in terms of the more
familiar μ_i (Reid and Beegle, 1971). For our purposes the 2N control
and state variables suffice for the definition of the thermodynamic
state at any point in the flow field.

There is very little that one can say about general relationships among the control variables and the state variables in systems removed from equilibrium. In some flows, however, the time scales characteristic of changes governed by the transport equations may be much longer than those characteristic of relaxation to thermodynamic equilibrium. For these systems it may be reasonable to approximate the behavior of thermodynamic quantities using established equilibrium relationships.

Yet, as Appendix B explains, more than one concept of equilibrium may apply. In a gradient-dynamic system, for example, there are at least two meaningful notions of equilibrium. One is that of stable equilibrium, in which the thermodynamic variables yield a local minimum in some postulated potential. The other is that of thermostatic equilibrium, in which the thermodynamic variables correspond to a global minimum in the potential. Hence thermostatic equilibria form a subclass of stable equilibria. Let us assume that the reservoir fluids in miscible gas floods behave very nearly as if they were locally in thermostatic equilibrium at each point in time and space.

This assumption justifies the use of classical thermostatics to model fluid-phase behavior. The mathematical conditions for stable equilibrium in a gradient-dynamic system imply an equation of state giving each point $(\eta_1,\ldots,\eta_{N-1}, -p)$ in equilibrium as a function of $(\omega_1,\ldots,\omega_{N-1}, V)$. At stable equilibria that are also thermostatic equilibria this equation of state renders a complete description of the thermodynamic system. However, for stable equilibria at which the global minimum in potential shifts from one local minimum to another several thermostatic equilibria coexist. Here the equation of state underdetermines the system. If there are two equiminima in potential, say, then there are two coexisting phases, V and L in our case. While both obey the equation of state, the equation alone yields no clue regarding their precise loci. The extra conditions determining the coexisting phases are, from Appendix B,

$$p_V = p_L$$

(2.1-1a)

$$\eta_i^V = \eta_i^L, \quad i = 1,\ldots,N-1$$

(2.1-1b)

In terms of the customary chemical potentials, (2.1-1b) is

$$\mu_i^V = \mu_i^L, \quad i = 1,\ldots,N-1$$

(2.1-1c)

These are the standard equations of equilibrium as stated by Gibbs.

Strictly speaking, the pressures of coexisting phases in a porous medium at thermostatic equilibrium differ owing to capillarity. Equations (2.1-1) fail to account for this phenomenon because the geometry of fluid phase boundaries does not appear as a control variable. In principle one can extend the present thermodynamics to include such effects using Gibbs' theory of capillarity (Gibbs, 1876 and 1878). However, there exist experimental data (Sigmund et al., 1973) indicating that capillary effects exert negligible influence on the distribution of species between coexisting fluids in typical hydrocarbon-saturated reservoir rocks. On the strength of this finding it appears reasonable to use equation (2.1-1) to define thermostatic equilibrium in the systems of interest here. In practice we shall use the pressure p_V in the vapor phase to compute equilibrium properties.

Geometry of equilibria.

This description of equilibria admits a geometric interpretation that has some heuristic value in the remainder of the chapter. Equations

(2.1-1) define the set of single-phase thermodynamic states that lie at
the limit of thermostatic equilibrium. This set is called the Maxwell
set K_M of the thermodynamic system. If a single-phase system confined
to thermostatic equilibrium crosses the Maxwell set, it bifurcates,
becoming a system of coexisting phases each of which lies on the Maxwell
set. Therefore in miscible gas floods the Maxwell set provides information, not only about when the fluids change from a single-phase regime
to a two-phase regime and vice versa, but also about the values of the
thermodynamic variables associated with each coexisting phase.

A similar demarcation exists for stability: the set of thermodynamic
states that lie at the limit of stable equilibrium is called the
spinodal set K_S of the system. Thermodynamic states that are stable
equilibria but are not thermostatic equilibria are metastable, meaning
that they may be observed under special circumstances but are labile.
Unstable points lying beyond the spinodal set are not observed. Arthur
S. Wightman, in his introduction to a monograph by Israel (1979),
reviews these physics more thoroughly. The primary object of this
chapter is to present schemes for computing the Maxwell set in simulators of miscible gas floods.

Critical points.

In miscible gas floods the critical points of the fluid mixtures play
an important role: they are points in the Maxwell set where coexisting
phases become indistinguishable. Therefore, mixtures in the critical
region flow very nearly as if the fluids were completely miscible.
Critical points are noteworthy, too, because in their vicinity standard
equation-of-state computations often perform poorly. For these reasons
it is useful to be able to compute critical points explicitly. Gibbs
(1876 and 1878, pp. 129-133) deduces two algebraic criteria for this
purpose. Appendix B reviews their derivation from the gradient-dynamic

viewpoint.

The first criterion for a point in thermostatic equilibrium to be a critical point is that it lie at the limit of convexity of the Helmholtz free energy with respect to the control variables. This is equivalent to requiring

$$
\det \begin{bmatrix} \partial_{\omega_1} \mu_1 & \cdots & \partial_{\omega_{N-1}} \mu_1 \\ & & \\ \cdot & & \\ \cdot & & \\ \cdot & & \\ \partial_{\omega_1} \mu_{N-1} & \cdots & \partial_{\omega_{N-1}} \mu_{N-1} \end{bmatrix} = U = 0
$$

$$(2.1\text{-}2)$$

The second criterion is that the critical point be a limit of points satisfying (2.1-2) for which isothermal variations in the control parameters can produce unstable phases. A necessary condition for this is

$$
\det \begin{bmatrix} \partial_{\omega_1} U & \cdots & \partial_{\omega_{N-1}} U \\ \partial_{\omega_1} \mu_2 & \cdots & \partial_{\omega_{N-1}} \mu_2 \\ & \cdot & \\ & \cdot & \\ & \cdot & \\ \partial_{\omega_1} \mu_{N-1} & \cdots & \partial_{\omega_{N-1}} \mu_{N-1} \end{bmatrix} = 0
$$

$$(2.1\text{-}3)$$

Equations (2.1-1) through (2.1-3) provide a complete description of the Maxwell set, provided we have a computable equation of state from which to calculate the chemical potentials.

2.2. Standard equation-of-state methods.

The theory summarized in the previous section provides both a geometric view and the equivalent algebraic conditions of local thermostatic equilibrium. Numerical simulation requires some computable form for this theory. Recent contributions to compositional reservoir simulation have shown a trend toward increasing reliance on equation-of-state methods to model fluid phase compositions and densities. These methods stand in contrast to approaches based on tabulated ratios $K_i = \omega_i^V/\omega_i^L$ of phase compositions for each species as functions of pressure and compositions.

The main advantage of equation-of-state methods is their thermodynamic consistency: the equations used to predict phase compositions are based on those used to predict phase densities, and thus the compatibility conditions (1.4-3) hold. Such consistency is necessary for the behavior of the thermodynamic system near critical points to vary smoothly with changes in pressure and mixture composition (Coats, 1980). This advantage is noteworthy, since, as Nolen (1973) explains, the convergence of the transport calculations near the miscible regions of gas floods is at stake.

However, the standard equation-of-state approaches have at least two disadvantages. First, the methods call for relatively powerful iterative techniques to solve for the fluid-phase characteristics that they predict. This mathematical machinery adds substantially to the cost of running compositional computer codes. Second, and more serious, the iterative techniques commonly used are rather sensitive to starting guesses near critical loci. This lack of reliability poses obstacles to the practical simulation of miscible gas floods. We shall elaborate on these observations in this section, saving Section 2.3 for the presentation of a simple approach to avoiding both difficulties.

The Peng-Robinson equation of state.

Peng and Robinson (1976) propose an equation of state for hydrocarbons that is cubic in molar volume V and is applicable to petroleum reservoir fluids through the use of mixing rules. Their work follows a long tradition of cubic equations of state beginning with that presented by van der Waals in 1873 and including the popular Redlich-Kwong equation (Redlich and Kwong, 1949). We shall use the Peng-Robinson equation in this investigation, although the general methodology we discuss applies to any similar cubic equation of state, including the Redlich-Kwong equation.

The Peng-Robinson equation is

$$p = RT/(V - b) - a(T)/[V(V + b) + b(V - b)]$$

(2.2-1)

where T is the temperature (K) and R is the gas constant, 8.31434 J/mol·K. The parameters a(T) and b are empirical factors calculated for pure substances according to rules that Peng and Robinson specify. Equation (2.2-1) is equivalent to

$$Z^3 - (1 - B)Z^2 + (A - 3B^2 - 2B)Z - (AB - B^2 - B^3) = 0$$

(2.2-2)

where

$$A = ap/R^2T^2$$

(2.2-3a)

$$B = bp/RT$$

(2.2-3b)

and

$$Z = pV/RT$$

(2.2-3c)

is the compressibility factor. When one solves equation (2.2-2) for a particular fluid phase, one may find three real roots. In this case the choice among roots is as follows: if the phase in question is a vapor, select the largest root; if the phase is a liquid, pick the smallest positive root. In practice we can solve equation (2.2-2) for all of its roots using a Laguerre iteration method (Smith, 1967) available in code as the IMSL subroutine ZPOLR.

Although Peng and Robinson introduce their equation for pure substances, for which the parameters a and b depend on thermodynamic properties of the individual molecular species, the equation extends to single-phase mixtures through mixing rules. These rules give the values of a and b for a mixed phase α in terms of their pure-substance values and the composition of the phase as follows:

$$a^\alpha = \sum_{i=1}^{N} \sum_{j=1}^{N} \omega_i^\alpha \omega_j^\alpha (1 - \delta_{ij}) \sqrt{(a_i a_j)}$$

(2.2-4a)

$$b^\alpha = \sum_{i=1}^{N} \omega_i^\alpha b_i$$

(2.2-4b)

Here a_i and b_i are values of the parameters a and b for species i, and δ_{ij} is a "binary interaction parameter" determined by fitting the mixture equation to experimental data. Oellrich et al. (1981) have published an extensive collection of values for δ_{ij}.

Peng and Robinson also present the equation for species fugacity in a single-phase mixture. This equation is an essential ingredient in phase-behavior predictions. The equation gives the fugacity f_i^α of species i in phase α as

$$f_i^\alpha = p \ \omega_i^\alpha \ \phi_i^\alpha$$

$$(2.2\text{-}5a)$$

where ϕ_i^α is the dimensionless fugacity coefficient given by

$$\phi_i^\alpha = (Z_\alpha - B^\alpha)^{-1} \ \exp \ [B_i(Z_\alpha - 1)/B^\alpha]$$

$$\times \ \{[Z_\alpha + (1 + \sqrt{2})B^\alpha]/[Z_\alpha - (1 - \sqrt{2})B^\alpha]\}^{-n_i^\alpha}$$

$$(2.2\text{-}5b)$$

with

$$n_i^\alpha = [A^\alpha/(B^\alpha\sqrt{8})] \ [(2/a^\alpha) \ \sum_{j=1}^{N} \ \omega_j^\alpha \ (1 - \delta_{ij}) \ \sqrt{(a_i a_j)} - b_i/b^\alpha]$$

$$(2.2\text{-}5c)$$

Hence f_i^α has the dimensions of pressure. In these formulas A^α and B^α are gotten by substituting a^α and b^α into equations (2.2-3a) and (2.2-3b), respectively. The determination of phase equilibria requires finding zeros of systems of algebraic equations, each of which is a combination of fugacities computed from equations (2.1-5). Let us discuss two such calculations used in standard equation-of-state reservoir simulators.

Saturation pressure calculations.

The problem of determining saturation pressures in a multicomponent
fluid mixture is the following: given a single-phase fluid of
prescribed temperature T and molar composition $(\omega_1,\ldots,\omega_{N-1})$, determine
a pressure p^{sat} at which a second, nascent phase begins to appear. If
the existing phase, denoted by ε, is a vapor and the nascent phase, say
ν, is a liquid, then $p^{sat} = p^{dew}$, the dew pressure. If ε is a liquid
and ν is a vapor, then $p^{sat} = p^{bub}$, the bubble pressure. Figure 2-1
depicts p^{bub} and p^{dew} as functions of composition for a typical binary
mixture.

The utility of saturation pressure calculations in compositional
modeling consists in discriminating between the one-phase and two-phase
regions in the space of thermodynamic states. In many petroleum reservoirs amenable to miscible gas flooding the practically attainable
one-phase region corresponds to pressures greater than p^{sat}. For these
systems determining p^{sat} amounts to locating points on a dome-like
surface lying above the composition space. Figure 2-2 shows such a dome
for a hypothetical ternary mixture. Simulator-predicted N-tuples
$(\omega_1,\ldots,\omega_{N-1}, p)$ lying under the saturation-pressure dome represent
two-phase states, and for these it is necessary to calculate vapor-
liquid equilibria to determine the properties of the two coexisting
phases. Since the set of all points $(\omega_1,\ldots,\omega_{N-1}, p^{sat})$ lying on the
dome is the Maxwell set of the fluid mixture, coexisting vapors and
liquids for two-phase states lie on this dome.

Mathematically, the saturation pressure problem for an N-component
mixture is a nonlinear set of N + 1 algebraic equations. N of these are
the conditions on chemical potentials that characterize the Maxwell set:

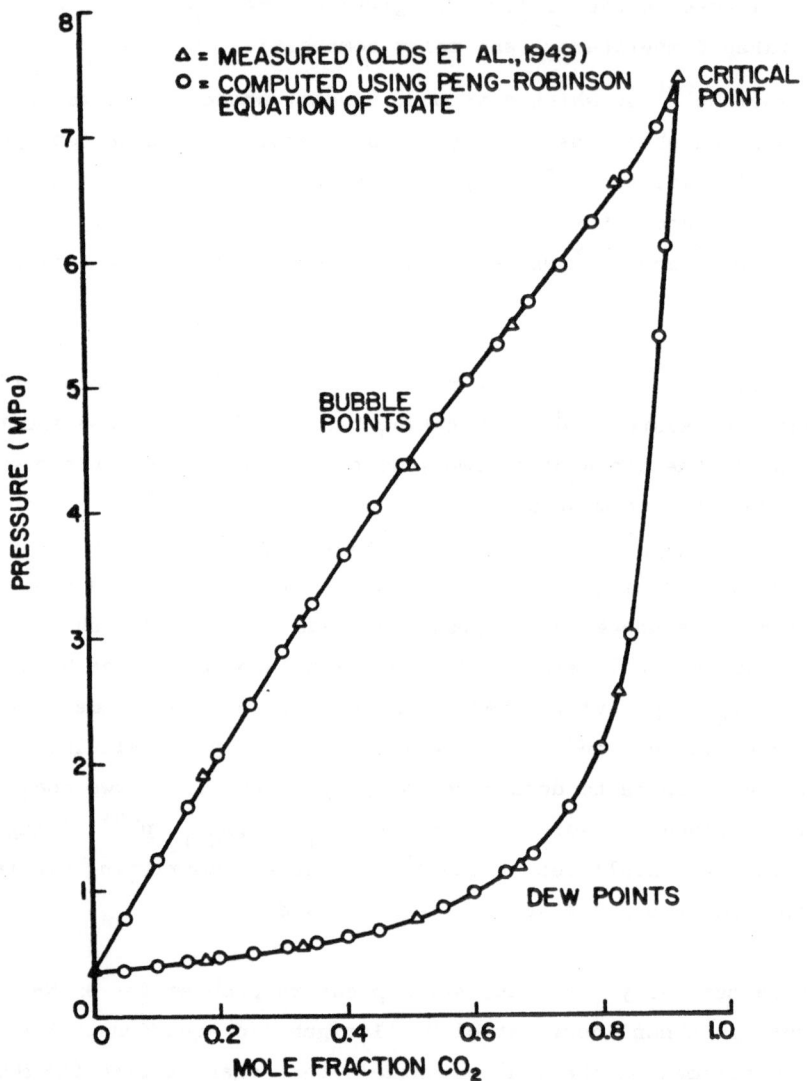

Figure 2-1. Saturation pressures for CO_2 + n-butane at 310.93K.

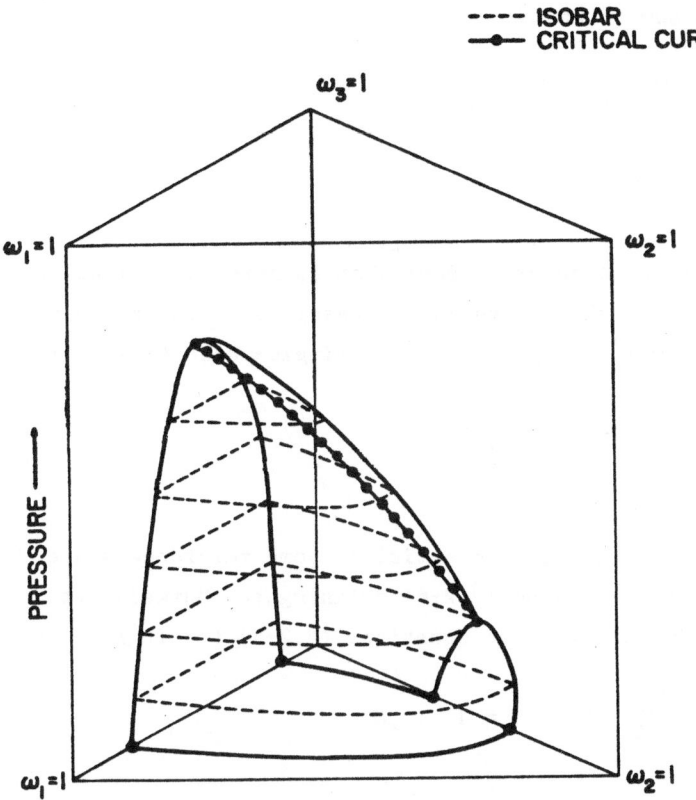

Figure 2-2. Saturation-pressure dome for a hypothetical ternary
            fluid mixture. Points under the dome belong to the
            two-phase region; points over the dome lie in the
            one-phase region. Coexisting phases lie on the dome.

$$\mu_i^\varepsilon = \mu_i^\nu, \quad i = 1, \ldots, N$$

$$(2.2\text{-}6a)$$

The last is the condition that mole fractions in the nascent phase sum to unity:

$$\sum_{i=1}^{N} \omega_i^\nu = 1$$

$$(2.2\text{-}6b)$$

Equations (2.2-6) assume a form that is more convenient computationally if we express them in terms of fugacities. The defining equation for the fugacity f_i^α of species i in a single-phase fluid α is

$$\mu_i^\alpha = RT \, \ell n \, f_i^\alpha + \mu_i^{ref}$$

$$(2.2\text{-}7)$$

where μ_i^{ref} is the chemical potential in some reference state. Now we can express the conditions (2.2-6) defining the Maxwell set in terms that we can compute from the equations (2.2-5) for fugacities:

$$f_i^\nu - f_i^\varepsilon = 0, \quad i = 1, \ldots, N$$

$$(2.2\text{-}8a)$$

$$p^{sat} - \sum_{i=1}^{N} f_i^\varepsilon / \phi_i^\nu = 0$$

$$(2.2\text{-}8b)$$

Here ϕ_i^α is the fugacity coefficient defined in equation (2.2-5a). For the problem at hand, the composition of the existing phase is known and the variables $(\omega_1^\nu, \ldots, \omega_N^\nu, p^{sat})$ are unknown. Therefore the system (2.2-8) has the following dependencies on the unknowns:

$$f_i^{\nu}(\omega_1^{\nu},\ldots,\omega_N^{\nu}, p^{sat}) - f_i^{\epsilon}(p^{sat}) = 0$$

$$\text{(2.2-9a)}$$

for $i = 1,\ldots,N$, and

$$p^{sat} - \sum_{i=1}^{N} f_i^{\epsilon}(p^{sat}) / \phi_i^{\nu}(\omega_1^{\nu},\ldots,\omega_{N-1}^{\nu}, p^{sat}) = 0$$

$$\text{(2.2-9b)}$$

Solving this system numerically requires an iterative method. The Newton-Raphson method is attractive except for the unwieldy computations needed to evaluate the derivatives of f_i^{ν} appearing in the Jacobian matrix. To avoid these computations, we can solve (2.2-9) using a quasi-Newton or secant method based on the use of finite differences.

For brevity, let us denote the algebraic system (2.2-9) by $R(\Phi) = 0$, where $\tilde{R}$: $R^{N+1} \to R^{N+1}$ is the nonlinear function whose roots we seek. Given an iterate $\tilde{\Phi}^k = (\omega_1^k,\ldots,\omega_N^k, p^{sat,k})$, the secant method computes a correction $\Delta\tilde{\Phi}^{k+1} = \tilde{\Phi}^{k+1} - \tilde{\Phi}^k$ according to the rule

$$J(\tilde{\Phi}^k) \, \Delta\tilde{\Phi}^{k+1} = -\tilde{R}(\tilde{\Phi}^k) = -\tilde{R}^k$$

$$\text{(2.2-10)}$$

Here J is a finite-difference approximation to the Jacobian matrix $\partial\tilde{R}/\partial\tilde{\Phi}$:

$$J_{ij}(\tilde{\Phi}^k) = [R_i(\tilde{\Phi}^k + h_j^k \, \tilde{e}_j) - R_i(\tilde{\Phi}^k)]/h_j^k$$

$$\text{(2.2-11)}$$

$\tilde{e}_j$ being the j-th unit basis vector. To start the algorithm, set $h_j^0 = c\Phi_j^0$ for some small factor $c < 1$, and thereafter choose $h_j^k = \Delta\Phi_j^k$. This choice yields an iterative procedure with a theoretical asymptotic

convergence rate (Ortega and Rheinboldt, 1970) of $(1 + \sqrt{5})/2 \simeq 1.618$.
However, plotting $\ln \|\tilde{R}^{k+1}\|$ versus $\ln \|\tilde{R}^{k}\|$ does not always give a line
with slope 1.618 in practice, although Figure 2-3 suggests that the
convergence rate is at least superlinear. The scheme requires $N + 2$
evaluations of the residual $\tilde{R}$, or $(N + 1)^2 + 1$ evaluations of fugacity
differences, at each iteration.

A simple damping routine increases the likelihood that the algorithm
will converge for poor initial guesses. This routine consists of
halving the correction vector repeatedly, if necessary, until $-0.1 \leq$
$(\omega_i^\nu)^k \leq 1.1$, $i = 1, \ldots, N$, aborting the iteration if a reasonable number
of halvings (say, 15) fails to give acceptable corrections.

The coded version of this algorithm performs well except near the
critical points of fluid mixtures, where the method is very sensitive to
initial guesses, $\tilde{\phi}^0$. This sensitivity reflects the proximity of true
roots of the system to "trivial roots", namely $\nu = \varepsilon$. Figure 2-4 plots
the progress of near-critical calculations for the binary mixture $CO_2 +$
n-butane at 310.93 K. Increasing sensitivity to initial guesses in the
critical region is a difficulty characteristic of standard equation-of-
state methods for computing saturation pressures and other phase
equilibria, and a considerable amount of research has focussed on
mitigating this sensitivity (see, for examples, Asselineau et al., 1979;
Baker and Luks, 1980; Gundersen, 1982; Nghiem and Aziz, 1979; Poling et
al., 1981; Risnes et al., 1981; Varotsis et al., 1981).

Table 2-1 compares saturation pressures for the binary mixture $CO_2 +$
n-butane as predicted by the Peng-Robinson equation with those measured
by Olds et al. (1949). In most cases the two sets of data agree fairly
well, although there are several exceptional points. Table 2-2 lists
the starting values and number of iterations required to achieve a

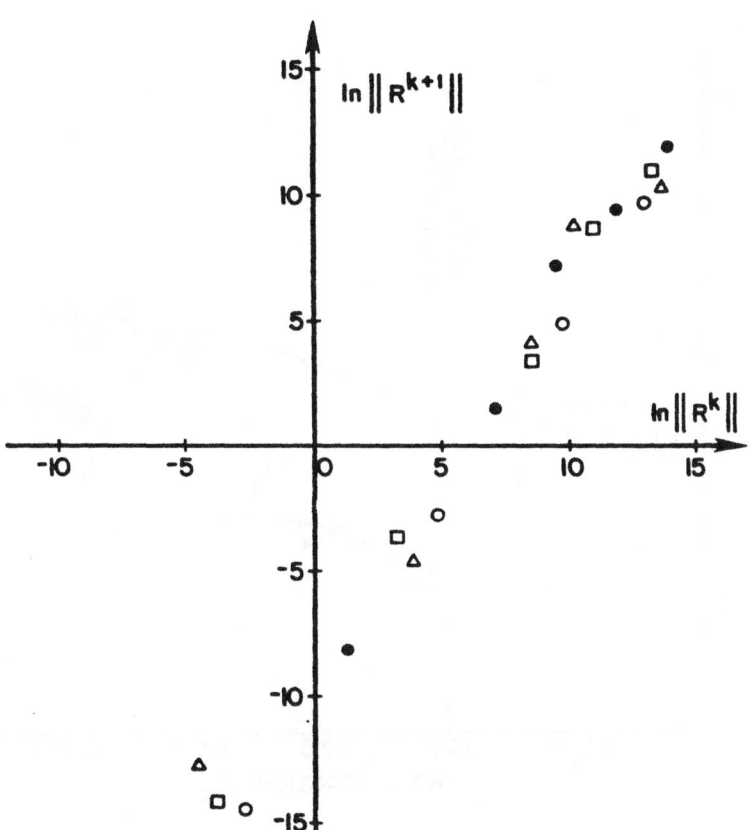

Figure 2-3. Plot of $\ln||\underset{\sim}{R}^{k+1}||$ vs. $\ln||\underset{\sim}{R}^k||$ for several saturation pressure calculations. The slopes for the lines of least-squares fit average 1.49.

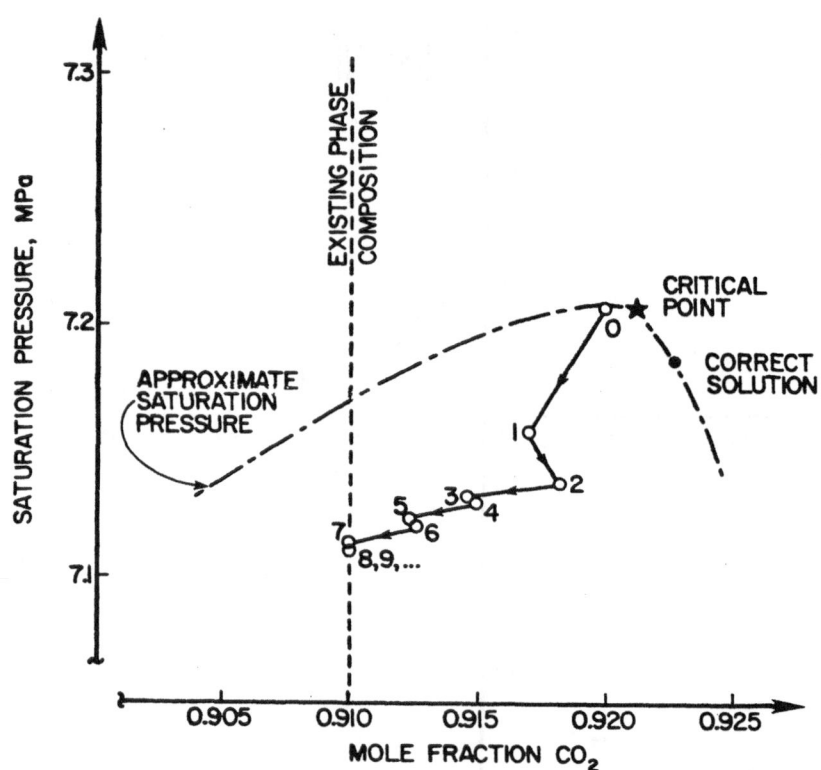

Figure 2-4.   Convergence of saturation pressure calculations to
a trivial root for CO_2 + n-balance at 310.93K using
the Peng-Robinson equation of state.

TABLE 2-1. COMPARISON OF CALCULATED SATURATION PRESSURES FOR CO_2 + n-BUTANE WITH DATA MEASURED BY OLDS ET AL. (1949).

Mole fraction CO_2	Temp. (K)	Dew pressure (MPa)			Bubble pressure (MPa)		
		Measured	Computed	Δ%	Measured	Computed	Δ%
0.8609	310.88	3.304	3.313	0.28	6.894	6.745	-2.12
	316.49	4.023	4.165	3.53	7.613	7.257	-4.67
0.7102	322.44	1.956	1.964	0.41	6.831	6.703	-1.87
	339.27	3.213	3.304	2.83	8.116	7.855	-3.22
0.6073	311.04	0.9988	0.9858	-1.30	5.050	5.086	0.71
	344.55	2.590	2.572	-0.68	7.613	7.432	-2.38
0.4551	328.16	1.125	1.119	-0.57	5.085	5.062	-0.45
	351.38	2.054	2.036	-0.85	6.503	6.352	-2.32
0.3671	322.16	0.8242	0.8051	-2.31	4.037	3.953	-2.09
	344.55	1.467	1.436	-2.12	5.253	5.056	-3.74
	366.49	2.480	2.421	-2.40	6.398	6.049	-5.45
0.1393	328.11	0.5728	0.6715	17.23	1.914	1.996	4.28
	349.94	1.006	1.136	12.92	2.535	2.635	3.94
	383.44	2.305	2.305	—	3.932	3.797	-3.43

TABLE 2-2. STARTING VALUES AND NUMBER k_{max} OF ITERATIONS TO CONVERGENCE
($\|R\|_2 < 10^{-2}$ Pa) FOR THE RUNS LISTED IN TABLE 2-1.

Mole fraction CO_2	Temp. (K)	Dew points, (MPa, mole fraction) Initial guess	Final value	k_{max}	Bubble points, (MPa, mole fraction) Initial guess	Final value	k_{max}
0.8609	310.88	(3.0,0.3)	(3.313,0.3547)	4	(6.0,0.9)	(6.745,0.9207)	11
	316.49	(4.0,0.4)	(4.165,0.4277)	4	(7.3,0.9)	(7.257,0.8916)	6
0.7102	322.44	(2.0,0.1)	(1.964,0.1515)	4	(6.0,0.8)	(6.703,0.8610)	7
	339.27	(3.0,0.2)	(3.304,0.2287)	4	(8.0,0.8)	(7.855,0.7721)	7
0.6073	311.04	(1.0,0.1)	(0.986,0.0717)	3	(4.0,0.9)	(5.086,0.8968)	5
	344.05	(2.0,0.1)	(2.572,0.1483)	5	(7.0,0.8)	(7.432,0.7651)	6
0.4551	328.16	(1.0,0.1)	(1.119,0.0534)	4	(4.0,0.8)	(5.062,0.8299)	5
	351.38	(2.0,0.1)	(2.036,0.0864)	3	(6.0,0.7)	(6.352,0.7328)	4
0.3671	322.16	(1.0,0.1)	(0.805,0.0325)	4	(3.0,0.8)	(3.953,0.8316)	4
	344.55	(1.0,0.1)	(1.436,0.0508)	5	(4.0,0.7)	(5.056,0.7493)	5
	366.49	(2.0,0.1)	(2.421,0.0811)	4	(5.0,0.6)	(6.049,0.6360)	5
0.1393	328.11	(1.0,0.1)	(0.672,0.0010)	5	(1.0,0.6)	(1.996,0.6703)	5
	349.94	(1.0,0.1)	(1.136,0.0152)	4	(1.0,0.5)	(2.635,0.5680)	8
	3 83.44	(2.0,0.1)	(2.305,0.0308)	5	(3.0,0.3)	(3.797,0.3869)	5

residual whose Euclidean norm

$$\|\bar{\underset{\sim}{R}}(\underset{\sim}{\Phi}^k)\|_2 = \{ \sum_{i=1}^{N} [\bar{R}_i(\underset{\sim}{\Phi}^k)]^2 \}^{\frac{1}{2}}$$

(2.2-12)

is less than 0.01 Pa for the runs reported in Table 2-1. Figure 2-5
shows plots of predicted and measured saturation pressures versus composition for the binary mixtures CO_2 + n-butane and CO_2 + n-decane at
344.26 K using experimental data from Olds et al. (1949) and Reamer and
Sage (1963). Figure 2-6 is a perspective plot of predicted saturation
pressures versus composition for the ternary mixture CO_2 + n-butane +
n-decane at 344.26 K.

Flash calculations of vapor-liquid equilibrium.

Given an isothermal fluid mixture $(\omega_1, \ldots, \omega_{N-1}, p)$ lying in the
two-phase region, it is necessary to determine the compositions
$(\omega_1^V, \ldots, \omega_N^V)$ and $(\omega_1^L, \ldots, \omega_N^L)$ of the coexisting phases and the mole
fractions Y_V and Y_L that they occupy in the mixture. In geometric
terms, this means locating the points on the Maxwell set that represent
coexisting states for the given feed $(\omega_1, \ldots, \omega_{N-1}, p)$ and calculating
their relative distances from the Maxwell set, as drawn in Figure 2-7.

Mathematically, this "flash" calculation amounts to solving a set of
algebraic equations for the 2N + 2 unknowns $\{\omega_1^\alpha, \ldots, \omega_N^\alpha, Y_\alpha, \alpha = V, L\}$.
The first three equations are the restrictions

$$Y_V + Y_L = 1$$

(2.2-13a)

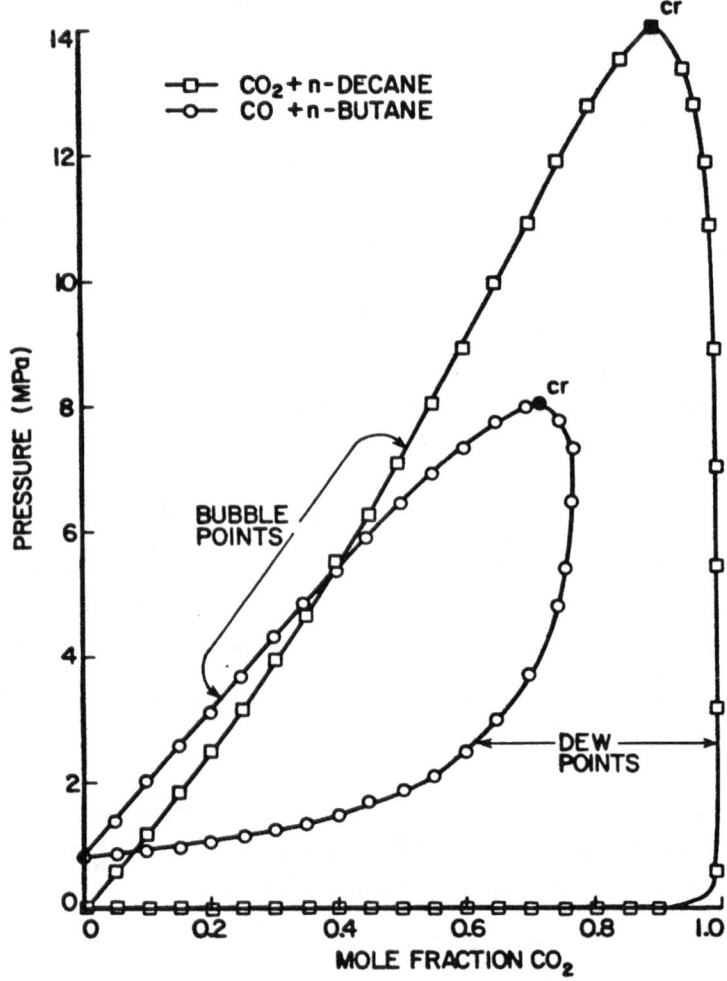

Figure 2-5. Saturation pressures for CO_2 + Hydrocarbon binaries
at 344.26K (=160F), computed using Peng-Robinson
equation of state.

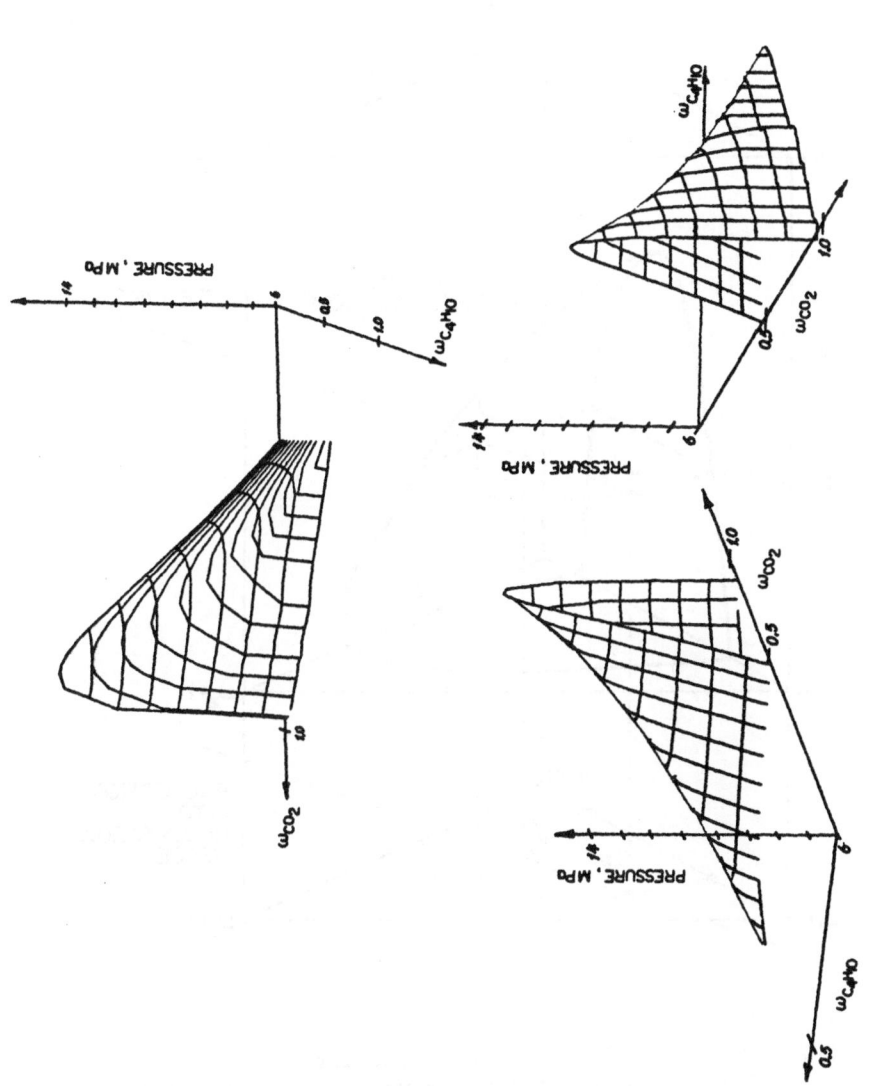

Figure 2-6. Three views of a perspective plot of the saturation-pressure dome for CO_2 + n-butane + n-decane at 344.26K, computed using the Peng-Robinson equation of state.

62

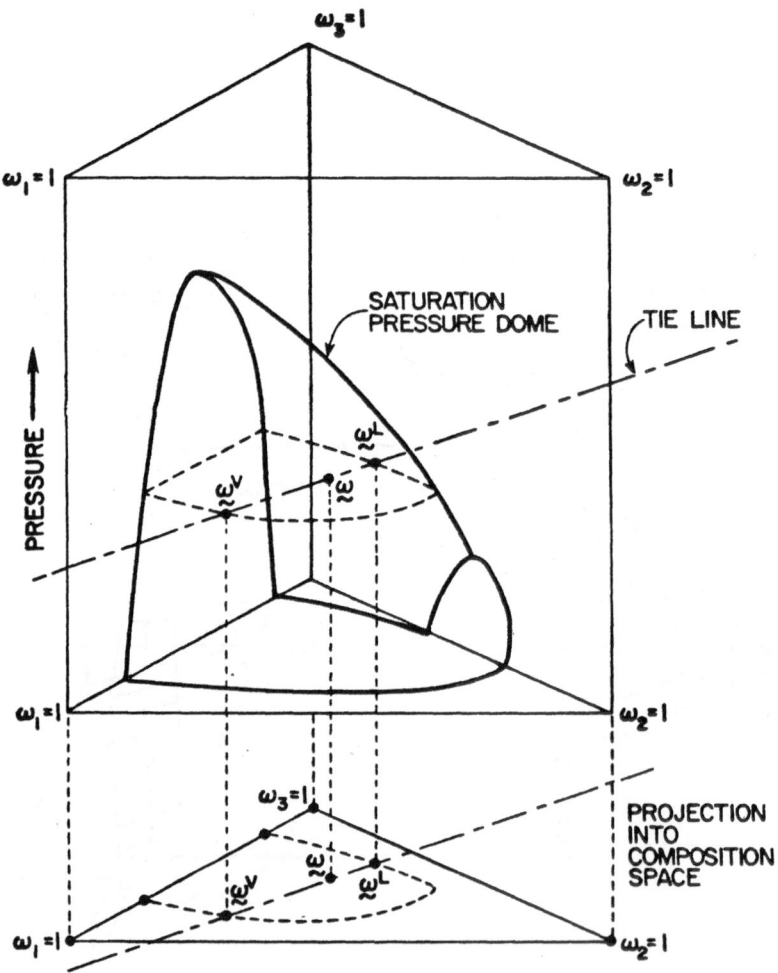

Figure 2-7. Geometry of vapor-liquid equilibria for a
hypothetical ternary mixture. The coexisting
phases $\underline{\omega}^V$ and $\underline{\omega}^L$ lying on the saturation dome
correspond to intersections of the tie line
passing through the feed composition $\underline{\omega}$ with the dome.

$$\sum_{i=1}^{N} \omega_i^\alpha = 1, \quad \alpha = V, L$$

$$(2.2\text{-}13b)$$

Next, we have N - 1 mole balances

$$Y_V \, \omega_i^V + Y_L \, \omega_i^L = \omega_i, \quad i = 1, \ldots, N\text{-}1$$

$$(2.2\text{-}13c)$$

Finally, there are N conditions of local thermostatic equilibrium,

$$R_i = f_i^V(\omega_1^V, \ldots, \omega_{N-1}^V, p) - f_i^L(\omega_1^L, \ldots, \omega_{N-1}^L, p) = 0$$

$$i = 1, \ldots, N$$

$$(2.2\text{-}13d)$$

The observation that equations (2.2-13a) through (2.2-13c) are fairly simple, while the conditions (2.2-13d) are complicated, motivates the following overall solution scheme, first described by Fussell and Yanosik (1978):

(i) Select N principal iteration variables, $\{\omega_i^L, \ldots, \omega_{N-1}^L, Y_L\}$ or $\{\omega_i^V, \ldots, \omega_{N-1}^V, Y_V\}$

(ii) Solve equations (2.2-13a) through (2.2-13c).

(iii) Compute the fugacity residuals $\bar{R}_i^k = (f_i^\alpha - f_i^\beta)^k$, i = 1,...,N, where the values of α and β depend on the choice in step (i).

(iv) If the residuals are sufficiently small, stop.

(v)   Correct the iteration variables using a quasi-Newton step.

(vi) Go to step (ii).

We shall discuss the implementation of these computations individually.

Step (i) poses the choice between two sets of principal iteration variables. These variables serve as the unknowns $\underset{\sim}{\Phi}$ in the quasi-Newton iteration of step (v). Let us choose

$$\underset{\sim}{\Phi} = \begin{cases} (\omega_1^L, \ldots, \omega_{N-1}^L, Y^L) & \text{if } Y_V \geq 0.5 \\[2ex] (\omega_1^V, \ldots, \omega_{N-1}^V, Y^V) & \text{if } Y_V < 0.5 \end{cases}$$

$$(2.2\text{-}14)$$

Fussell and Yanosik (1978) report that this choice leads to more reliable convergence of the scheme than arbitrary selection of one set of variables for $\underset{\sim}{\Phi}$. For the rest of this section let us identify the phase chosen for the iteration variables by the index α.

Step (ii) calls for the solution of equations (2.2-13a) through (2.2-13c). This is a straightforward matter given the iterates $\{\omega_i^\alpha, \ldots, \omega_{N-1}^\alpha, Y_\alpha\}$: from equations (2.2-13),

$$Y_\beta^k = 1 - Y_\alpha^k$$

$$(2.2\text{-}13a')$$

$$\omega_i^{\beta,k} = (\omega_1 - Y_\alpha^k \omega_i^{\alpha,k})/Y_\alpha^k, \qquad i = 1, \ldots, N$$

$$(2.2\text{-}13c')$$

$$\omega_N^{\alpha,k} = 1 - \sum_{i<N} \omega_i^{\alpha,k}$$

$$(2.2\text{-}13b')$$

$$\omega_N^{\beta,k} = 1 - \sum_{i<N} \omega_i^{\beta,k}$$

$$(2.2\text{-}13b'')$$

Once these values are available, one can compute the fugacity residuals
(step (iii)) as functions of the principal iteration variables by the
definition of the R_i implied in equation (2.2-13d). To check the
residuals, as prescribed in step (iv), we can compute their Euclidean
norm and test it against a numerical convergence criterion, usually no
smaller than 0.01 Pa.

The quasi-Newton iteration required in step (v) is similar to that
used in the saturation pressure calculations. To solve the nonlinear
system $\bar{R}(\Phi) = 0$, let us compute a correction vector $\Delta\Phi^{k+1} = \Phi^{k+1} - \Phi^{k}$ at
each iteration k by solving the linear system

$$J_{ij}(\Phi^k) \, \Delta\Phi_j^{k+1} = - \bar{R}_i(\Phi^k) = - \bar{R}_i^k$$

$$(2.2-15)$$

Here, as in equation (2.2-10), the matrix elements J_{ij} are finite-
difference approximations to elements $\partial\bar{R}_i/\partial\Phi_j$ of the Jacobian matrix:

$$J_{ij}(\Phi^k) = [\bar{R}_i(\Phi^k + h_j^k e_j) - \bar{R}_i(\Phi^k)]/h_j^k$$

$$(2.2-16)$$

After picking initial values for h_j^0 we choose the subsequent increments
h_j^k according to the secant rule,

$$h_j^k = \max \{\Delta\Phi_j^k, \, h_j^0\}, \, k > 0.$$

$$(2.2-17)$$

Putting a lower bound on the increments in this way guards against
division by zero in case one component is absent and the corresponding
correction vanishes.

TABLE 2-3. SUMMARY OF FLASH CALCULATIONS FOR CO_2 + n-BUTANE + n-DECANE
AT 344.26 K USING THE PENG-ROBINSON EQUATION OF STATE

Pressure (MPa)	Feed composition*		Vapor composition*		Liquid composition*		Y_V
	CO_2	$n-C_4H_{10}$	CO_2	$n-C_4H_{10}$	CO_2	$n-C_4H_{10}$	
7.0	0.5900	0.0350	0.9842	0.0123	0.4912	0.0407	0.2004
7.0	0.5900	0.0754	0.9694	0.0272	0.4896	0.0882	0.2092
7.0	0.5900	0.1731	0.9299	0.0669	0.4481	0.2049	0.2307
7.0	0.5900	0.2811	0.8772	0.1201	0.4946	0.3346	0.2493
7.0	0.5900	0.3300	0.8476	0.1501	0.5046	0.3897	0.2490
8.0	0.6700	0.0350	0.9819	0.0136	0.5486	0.0433	0.2802
8.0	0.6700	0.0754	0.9651	0.0305	0.5485	0.0939	0.2916
8.0	0.6700	0.1731	0.9178	0.0780	0.5543	0.2175	0.3183
8.0	0.6700	0.2268	0.8852	0.1108	0.5651	0.2833	0.3277
8.0	0.6700	0.2811	0.8414	0.1552	0.5917	0.3386	0.3134
9.0	0.7189	0.0350	0.9792	0.0148	0.6045	0.4386	0.3053
9.0	0.7189	0.0754	0.9604	0.0336	0.6071	0.0948	0.3164
9.0	0.7189	0.1220	0.9358	0.0580	0.6133	0.1532	0.3274
9.0	0.7189	0.1731	0.9033	0.0904	0.6274	0.2141	0.3315
9.0	0.7189	0.2268	0.8554	0.1379	0.6631	0.2631	0.2902
10.0	0.7617	0.0200	0.9824	0.0092	0.6579	0.2508	0.3196
10.0	0.7617	0.0349	0.9750	0.0614	0.6598	0.0438	0.3232

*Mole fractions.

TABLE 2-3. (CONTINUED)

Pressure (MPa)	Feed composition* CO$_2$	Feed composition* n-C$_4$H$_{10}$	Vapor composition* CO$_2$	Vapor composition* n-C$_4$H$_{10}$	Liquid composition* CO$_2$	Liquid composition* n-C$_4$H$_{10}$	Y$_v$
10.0	0.7617	0.0741	0.9531	0.0379	0.6671	0.0939	0.3305
10.0	0.7617	0.1220	0.9222	0.0678	0.6832	0.1485	0.3282
10.0	0.7616	0.1731	0.8728	0.1145	0.7244	0.1927	0.2509
11.0	0.8008	0.0200	0.9768	0.0104	0.7119	0.0248	0.3358
11.0	0.8008	0.0350	0.9678	0.0188	0.7159	0.0423	0.3370
11.0	0.8008	0.0754	0.9394	0.0449	0.7329	0.0903	0.3289
11.0	0.8008	0.1000	0.9169	0.0648	0.7507	0.1152	0.3014
11.0	0.8008	0.1220	0.8831	0.0884	0.7785	0.1306	0.2038
12.0	0.8345	0.0200	0.9655	0.0122	0.7675	0.0240	0.3385
12.0	0.8345	0.0350	0.9528	0.0225	0.7762	0.0411	0.3301
12.0	0.8345	0.0500	0.9378	0.0342	0.7879	0.0571	0.3108
12.0	0.8345	0.0600	0.9254	0.0430	0.7985	0.0667	0.2835
13.0	0.8610	0.0100	0.9509	0.0072	0.8197	0.0113	0.3144
13.0	0.8610	0.0200	0.9730	0.0153	0.8307	0.0219	0.2849
13.0	0.8610	0.0300	0.9176	0.0247	0.8473	0.0313	0.1954

*Mole fractions

Just as the saturation pressure scheme described earlier, this algorithm has an asymptotic convergence rate of about 1.618, although the caveats regarding plots of $\ln \|\underline{\bar{R}}^{k+1}\|$ versus $\ln \|\underline{\bar{R}}^{k}\|$ apply here as well. Table 2-3 summarizes some results of flash calculations for the ternary mixture CO_2 + n-butane + n-decane at 344.26 K.

This algorithm requires $N^2 + 1$ evaluations of fugacity differences per iteration. It also exhibits the same sorts of difficulties as that for the saturation pressures: it becomes increasingly sensitive to initial guesses near critical points, where the compositions of the coexisting phases become identical. In addition to the investigators cited for saturation pressure calculations, Kao (1978), Li and Nghiem (1982), Mehra et al. (1982), and Michelsen (1980) have discussed various strategies aimed at improving starting guesses and narrowing the regions of poor convergence.

Critical point calculations.

Given an equation of state it is possible to compute critical points of fluid mixtures. Although such calculations are not always necessary in the standard equation-of-state approach to reservoir simulation, they are important in constructing the Maxwell set interpolation scheme described in Section 2.3. The problem of computing critical points for an isothermal mixture may be simply stated as follows: given values $(\omega_3, \ldots, \omega_{N-1}, p)$, determine the point (ω_1, ω_2) along the critical curve, that is, along the locus of points at which the critical conditions (2.1-15) and (2.1-16) hold. Although Gibbs (1876 and 1878, pp. 129-133) formulated this problem over a century ago, reports of actual computational experience have been relatively recent (Peng and Robinson, 1977; Baker and Luks, 1980; Heideman and Khalil, 1980; Peng and Robinson, 1980).

The equations to be solved in critical point calculations are equations (2.1-15) and (2.1-16), which by the definition (2.2-7) of fugacity are equivalent to

$$U = RT \det \begin{bmatrix} (\partial_{\omega_1} f_1)/f_1 & \cdots & (\partial_{\omega_{N-1}} f_1)/f_1 \\ & & \\ \cdot & & \\ \cdot & & \\ \cdot & & \\ (\partial_{\omega_1} f_{N-1})/f_{N-1} & \cdots & (\partial_{\omega_{N-1}} f_{N-1})/f_{N-1} \end{bmatrix} = 0$$

$$(2.2-18a)$$

and a similiar equation gotten from (2.2-18a) by replacing the first row in the matrix with the vector

$$(\ (\partial_{\omega_1} U)/RT, \ \ldots, \ (\partial_{\omega_{N-1}} U)/RT \ )$$

$$(2.2-18b)$$

as stipulated in Appendix B. For a three-component system these reduce to

$$U = RT \ [(\partial_{\omega_1} f_1)(\partial_{\omega_2} f_2) - (\partial_{\omega_2} f_1)(\partial_{\omega_1} f_2)]/(f_1 f_2) = 0$$

$$(2.2-19a)$$

$$RT \ [(\partial_{\omega_1} U)(\partial_{\omega_2} f_2) - (\partial_{\omega_2} U)(\partial_{\omega_1} f_2)]/f_2 = 0$$

$$(2.2-19b)$$

In this case, the derivatives of U with respect to ω_i are

$$\partial_{\omega_i} U = \{- U [f_2(\partial_{\omega_i} f_1) + f_1(\partial_{\omega_i} f_2)] + RT [(\partial_{\omega_1} \partial_{\omega_i} f_1) (\partial_{\omega_i} f_2)$$

$$+ (\partial_{\omega_1} f_1) (\partial_{\omega_2} \partial_{\omega_i} f_2) - (\partial_{\omega_1} \partial_{\omega_i} f_1) (\partial_{\omega_1} f_2)$$

$$- (\partial_{\omega_2} f_1) (\partial_{\omega_1} \partial_{\omega_i} f_2)]\} / (f_1 f_2) = 0$$

$$(2.2-20)$$

for i = 1 or 2.

Although one could in principle compute by analytic formulas the various derivatives of fugacity appearing in these equations, the task is tedious at best for first derivatives and worse for second derivatives. Moreover, from a numerical standpoint the resulting analytic expressions are complicated combinations of transcendental functions, and control over the truncation and roundoff errors associated with their literal transcription to Fortran is uncertain. One alternative is to use simple centered finite-difference approximations to these terms, for example,

$$\partial_{\omega_1} f_i \simeq [f_i(\omega_1 + \Delta\omega, \omega_2) - f_i(\omega_1 - \Delta\omega, \omega_2)]/(2\Delta\omega)$$

$$(2.2-21)$$

In practice, successively smaller values of $\Delta\omega$ lead to estimates of a given derivative that appear to converge quickly to a single value, until $\Delta\omega$ becomes so small that the resulting differences in f_i are comparable to the machine's limits on precision. For double-precision calculations on an IBM 3081 the value $\Delta\omega = 10^{-5}$ works well.

Using these finite-difference approximations, we can solve for the roots of the system (2.2-18) using the secant quasi-Newton algorithm published by Wolfe (1959) and available in coded version from IMSL as

the subroutine ZSCNT. Figure 2-8 shows points along the critical curve
computed for the ternary mixture CO_2 + n-butane + n-decane at 344.26 K.

Computational considerations.

As this section has suggested, the equation-of-state approach to
compositional reservoir simulation, at least as commonly implemented,
has some undesirable traits. To begin with, methods based on solving
equal-fugacity constraints pointwise require a fair degree of computational sophistication, and one effect of this fact is to divorce the
numerics from much of the global geometric picture of miscible gas flood
thermodynamics. This criticism is more esthetic than damning, but it
has some significance. Much of the current understanding of the design
principles for miscible gas floods rests on geometric pictures using the
Maxwell set; papers by Hutchinson and Braun (1961), Metcalfe et al.
(1973), Metcalfe and Yarborough (1979), and Orr and Jensen (1982) are
four among many examples of such work.

A second, more salient consequence of the computational sophistication required in the standard approaches is their cost. These
approaches entail saturation pressure calculations or flash calculations
or both for each time step, for each spatial node, and for each iteration in a transport code. This overhead is expensive, and it hinders
the application of compositional simulation to large-scale studies.
Moreover, the expense of solving equal-fugacity constraints repeatedly
during simulation is incommensurate with the quality of the results they
predict. Cubic equations of state are inherently limited in the
accuracy with which they model fluid-phase behavior (Abbott, 1979). A
cheaper prediction method giving results of comparable veracity and
thermodynamic consistency would be more appropriate for use in simulators.

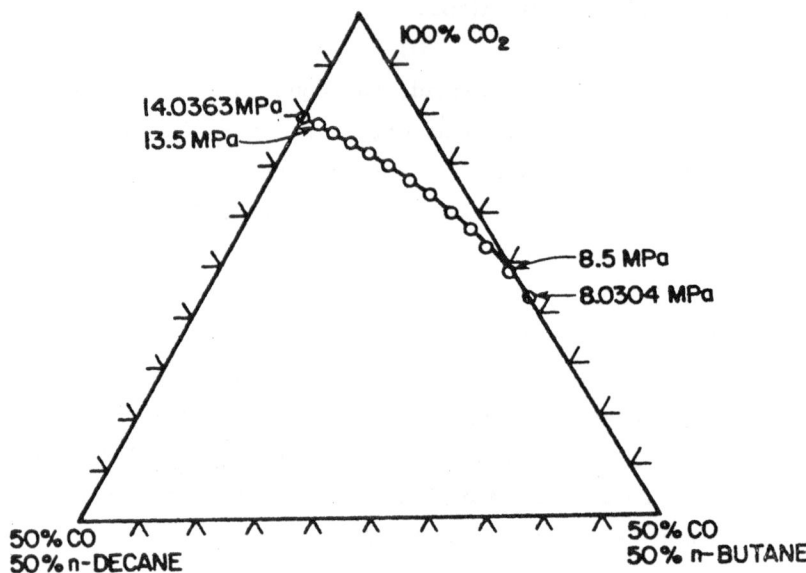

Figure 2-8. Projection of the critical curve onto composition
space for CO_2 + n-butane + n-decane at 344.26K,
computed using the Peng-Robinson equation of state.
Computed points lie at 0.5 MPa increments between the
binary endpoints.

The most serious difficulty with the standard equation-of-state methods is their unreliability. In the most naive applications the failure of thermodynamic calculations owing to inadequate starting guesses can cause a simulator to abort. One tactic for avoiding this is to identify the region of starting-guess sensitivity with the region of very low interfacial tensions, using failure to converge as prima facie evidence, as it were, for miscibility. This approach, while entirely practicable, lacks sound justification and carries the risk of thermodynamic inconsistency if ineffective starting guesses occur outside the critical region. Calculations in the critical region would be more believable if they converged.

2.3. Maxwell-set interpolation.

The shortcomings of the standard equation-of-state methods arise when
the unknown properties of the fluid mixture lie on the Maxwell set, for
which we have only an implicit representation. This observation
suggests circumventing the difficulties by developing an explicit representation of the Maxwell set whose numerical evaluation is cheaper and
less likely to fail. So long as this representation is reasonably
accurate and preserves thermodynamic consistency, we can avoid solving
for equal fugacities during the course of transport calculations while
still imposing equation-of-state constraints to compute fluid-phase
thermodynamics.

This section describes such an approach. Specifically, let us
consider calculating the saturation-pressure dome by interpolation,
using data generated by the standard methods of Section 2.2. In this
approach we can represent vapor-liquid equilibria by a set of tie lines
also based on the results of the standard methods. The data supporting
this scheme are computed prior to any flow simulation. One therefore
encounters difficulties associated with the standard techniques only in
the construction of a database, not in the midst of transport calculations. The interpolation technique affords order-of-magnitude or
greater reductions in the time required to execute two-phase calculations and virtually eliminates sensitivity to starting guesses. We
shall develop the interpolation scheme for a three-component system;
however, as Chapter Five discusses, the construction extends to larger
numbers of components.

The saturation-pressure dome.

One simple and easily generalized method for interpolating a real-valued function of two variables given data at discrete points of its domain is the method of plates. For the saturation-pressure dome of a three-component system, the domain is the triangular subset of R^2 defined by $\Omega_2 = \{(\omega_1, \omega_2) \in R^2 \mid \omega_1 + \omega_2 = 1 \}$. Let κ be a collection $\{(\omega_{1,k}, \omega_{2,k})\}_{k=1}^{K}$ of knots in this domain, and consider a proper triangulation Ω_2 generated by taking vertices from κ (see Prenter, 1975, Section 5.4). Given nodal values $p_k^{sat} = p^{sat}(\omega_{1,k}, \omega_{2,k})$ for all knots, the method of plates interpolates p^{sat} as

$$p^{sat}(\omega_1, \omega_2) = \sum_{k=1}^{K} p_k^{sat} T_k(\omega_1, \omega_2)$$

$$(2.3-1)$$

where $\{T_k\}_{k=1}^{K}$ is the basis giving p^{sat} as a plane over each subset of Ω_2 belonging to the triangularization. In practice one computes (2.3-1) for a given triangular region using area coordinates (Pinder and Gray, 1976, Section 4.8).

The triangular linear interpolation scheme (2.3-1) is easy to compute, and it extends readily to functions of several variables. In addition, if $p^{sat} \in C^2(\Omega_2)$ the interpolation error is subject to control, obeying

$$\|p^{sat} - p^{sat}\|_{\infty} \leq 4 M_2 h^2$$

$$(2.3-2)$$

where h denotes the mesh of the triangularization and M_2 is a bound on $|\partial^2 p^{sat}/\partial \omega_i \partial \omega_j|$ over Ω_2, $i,j \leq 2$ (Prenter, 1975, Section 5.4).

Several practical rules promote the construction of good interpolation schemes for the saturation pressure dome. Two of these are fairly
obvious. First, the dome need only be computed in regions of Ω_2 where
p^{sat} exceeds the pressures expected to be encountered in the transport
problem. We can therefore confine the interpolation scheme to these
regions when we know in advance the operating pressure range of
interest. Second, to facilitate the mechanical search for the grid
element containing specific arguments (ω_1, ω_2), it is helpful to choose
knots along lines of constant ω_1 and ω_2.

Furthermore, to retain the advantages of the equation-of-state
method, the interpolation scheme ideally should preserve thermodynamic
consistency. This implies in particular that the scheme should not give
interpolates of the Maxwell set that are thermodynamically unstable. In
the geometric terms of Section 2.1, the scheme should not produce values
that lie on the wrong side of the spinodal set K_S. This restriction
typically causes no concern except near critical loci, where the Maxwell
set K_M and the spinodal set K_S intersect as drawn in Figure 2-9. Here
an approximation $\hat{p}^{sat}$ that is satisfactory in terms of its absolute
error $|p^{sat} - \hat{p}^{sat}|$ can be inadmissible on thermodynamic grounds. One
way to avoid such anomalies is to force a more accurate representation
of the critical region by choosing points along the critical locus as
knots. This tactic calls for the critical-point calculations described
in Section 2.2.

Finally, in some cases interpolation using artificial data may
improve the scheme. In nature saturation pressure domes often have
regions of high curvature near one or more of the boundaries $\omega_i = 0$ of
the composition domain. For such surfaces the constant M_2 in the error
estimate (2.3-2) is large, and thus the interpolation error may also be
large. Figure 2-10 shows schematically how judicious choices of artifi-

77

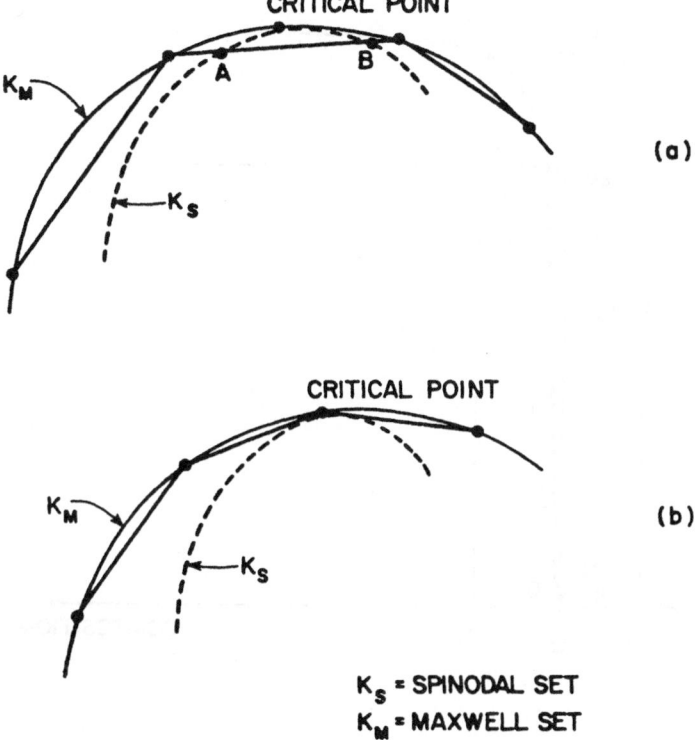

Figure 2-9. Thermodynamically inadmissible (a) and admissible (b) interpolation schemes for the Maxwell set. The segment AB in (a) is thermodynamically unstable.

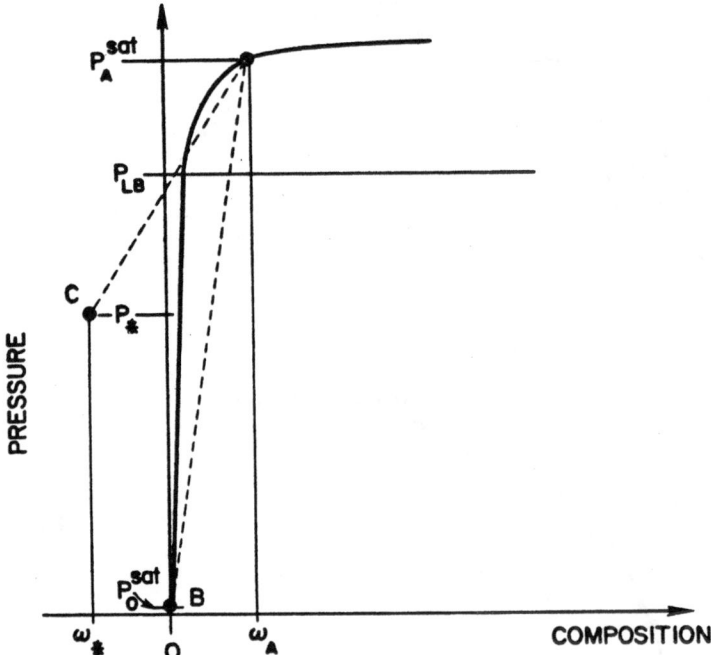

Figure 2-10. A cross-section of a hypothetical saturation
pressure dome showing how an artificial knot
ω_* can yield better interpolated values than
the true knot at 0.

cial knots can lead to better estimates of p^{sat} over the operating
pressure range in these cases. The figure depicts, in cross section,
part of a hypothetical p^{sat} dome, the horizontal axis being a line in
the composition domain Ω_2 and p_{LB} representing a lower bound on
pressures encountered during transport simulation. Given a knot at ω_A
with saturation pressure p_A^{sat}, a "real" knot at the composition value 0
with $\hat{p}^{sat} = p_0^{sat}$ leads to poor interpolated values on the composition
interval $(0, \omega_A)$ owing to the large curvature in p^{sat} there. In this
hypothetical case, choosing an artificial (in fact, non-physical) knot
at ω_* and assigning $\hat{p}^{sat}(\omega_*) = p_*$ gives better interpolated values so
long as $\hat{p}^{sat} \geq p_{LB}$. The larger errors occuring for $\hat{p}^{sat} < p_{LB}$ are not
important, since we compute the Maxwell set only when $\hat{p}^{sat} > p_{LB}$. The
addition of artificial knots may change the interpolation domain Ω_2 to a
somewhat different subset Ω_2^+ of R^2

Consider this scheme as implemented for the system CO_2 + n-butane +
n-decane at 344.26 K. This fluid mixture exhibits many of the qualitative features associated with the phase behavior of CO_2 - reservoir oil
mixtures, although the latter are vastly more complex in composition.
The CO_2 + n-butane + n-decane system also has the advantage of having
been studied as a model for CO_2 displacement mechanisms in miscible gas
flooding (Metcalfe and Yarborough, 1979; Orr and Jensen, 1982). Let the
species indices 1, 2, and 3 refer to CO_2, n-butane, and n-decane,
respectively.

Figure 2-11 shows the interpolation grid for the saturation-pressure
dome of this system. The grid exemplifies all four of the observations
discussed above. First, the grid covers only a portion of Ω_2. Since no
CO_2 flood of this system operating at pressures below the lowest
critical pressure can be miscible, there is no need to compute the
Maxwell set at values lying very far below this pressure. The lowest

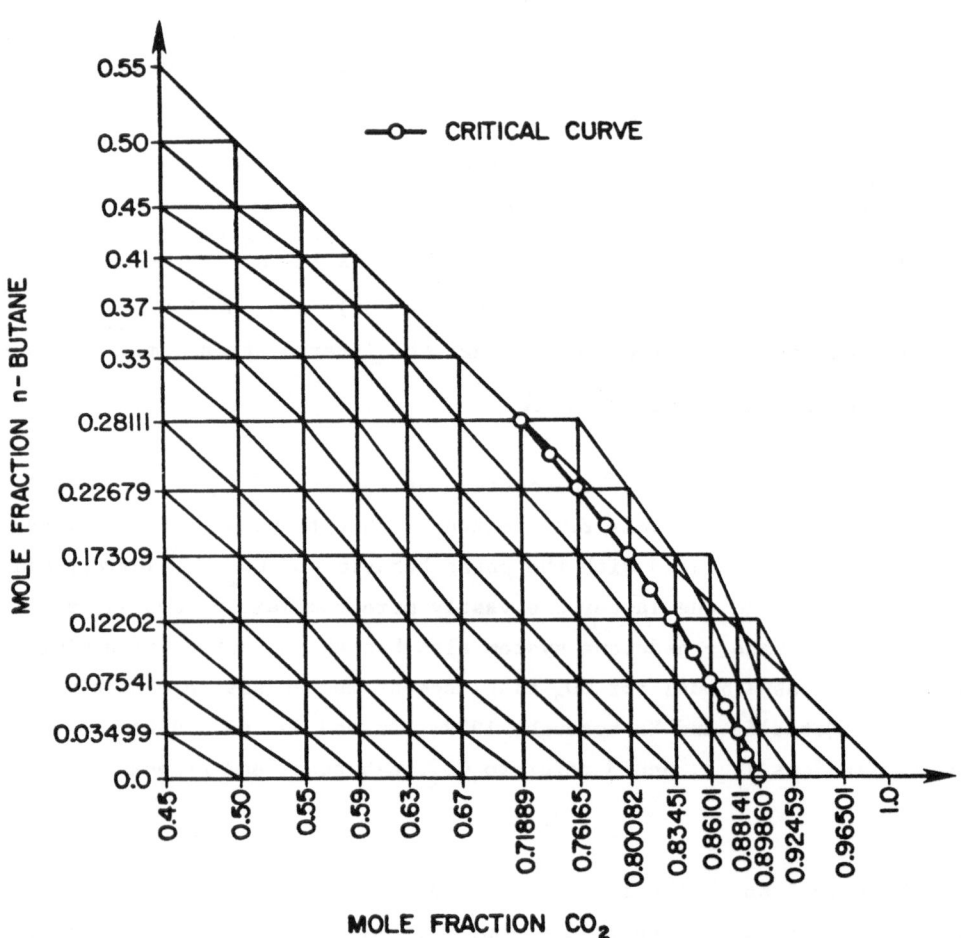

Figure 2-11. Interpolation grid for saturation pressures
of CO_2 + $n-C_4H_{10}$ + $n-C_{10}H_{22}$ at 344.26K.

pressure on the critical curve in this case is 8.030 MPa, occurring at $\omega_1 = 0.71889$, $\omega_2 = 0.28111$. Thus the interpolation scheme exists on a region Ω_2^+, excluding a large zone where $p^{sat} < 6.0$ MPa by including only knots where $\omega_1 > 0.45$. In keeping with our second observation, the knots of the interpolation scheme lie along lines of constant ω_1 and ω_2. Third, several critical points appear as knots, including the critical points for the binary mixtures CO_2 + n-butane and CO_2 + n-decane. In Figure 2-11 the critical knots appear as circles. Finally, the triangulation extends beyond Ω_2 to accommodate the artificial knots mentioned above. A service routine for triangular meshes (see, for example, Page, 1982) is handy in assigning values of p^{sat} to these knots by trial and error.

Using this interpolation scheme, computing saturation pressures is a simple matter. For a given composition (ω_1^0, ω_2^0) the algorithm requires a search through a list of grid elements, followed by evaluation of the basis functions associated with the vertices of the element to which (ω_1^0, ω_2^0) belongs. If the K knots lie along lines of constant ω_1 and ω_2, then the search need only be $O(\sqrt{K})$ in length. The interpolation itself requires less than 50 floating-point operations using area coordinates.

Vapor-liquid equilibria.

Given a geometric representation of the Maxwell set, it is possible to develop a geometric method for computing vapor-liquid equilibria. In particular, it is possible to construct a family of tie lines connecting points on the saturation-pressure dome that represent coexisting fluid phases. Then the fluid phases corresponding to a point (ω_1, ω_2, p) lying under the p^{sat} dome are just the points $(\omega_1^V, \omega_2^V, p)$, $(\omega_1^L, \omega_2^L, p)$ at which the appropriate tie line intersects the dome, as drawn in Figure

2-7. The mole fractions occupied by the phases vary linearly with distance from the dome:

$$Y_V = \left[\frac{(\omega_1^L - \omega_1)^2 + (\omega_2^L - \omega_2)^2}{(\omega_1^L - \omega_1^V)^2 + (\omega_2^L - \omega_2^V)^2} \right]^{\frac{1}{2}}$$

(2.3-3)

Hence the tie lines and the representation of the Maxwell set give all of the information necessary to determine vapor-liquid equilibria.

Suppose we represent the tie lines for the model system CO_2 + n-butane + n-decane at 344.26 K by interpolating data generated using the methods of Section 2.2. At a given pressure p, each tie line has the form

$$\omega_2 = \psi(p) \, \omega_1 + \omega^*(p)$$

(2.3-4)

where ω^* is the intercept with the axis $\omega_1 = 0$. According to a rule of thumb called Hand's rule (Hand, 1930; Van Quy et al., 1972), ω^* is a constant for each pressure. However, this rule can be unreliable for some systems, and it is better instead to assume that at each pressure ψ and ω^* vary with p but obey the approximate rule

$$\psi \simeq \hat{\psi} = \alpha(p) \, \omega^* + \beta(p)$$

(2.3-5)

Figure 2-12 plots these lines for various pressures. We can compute α and β at discrete pressures by linear least-squares fit, using linear interpolation to compute α and β at other pressures. The least-squares

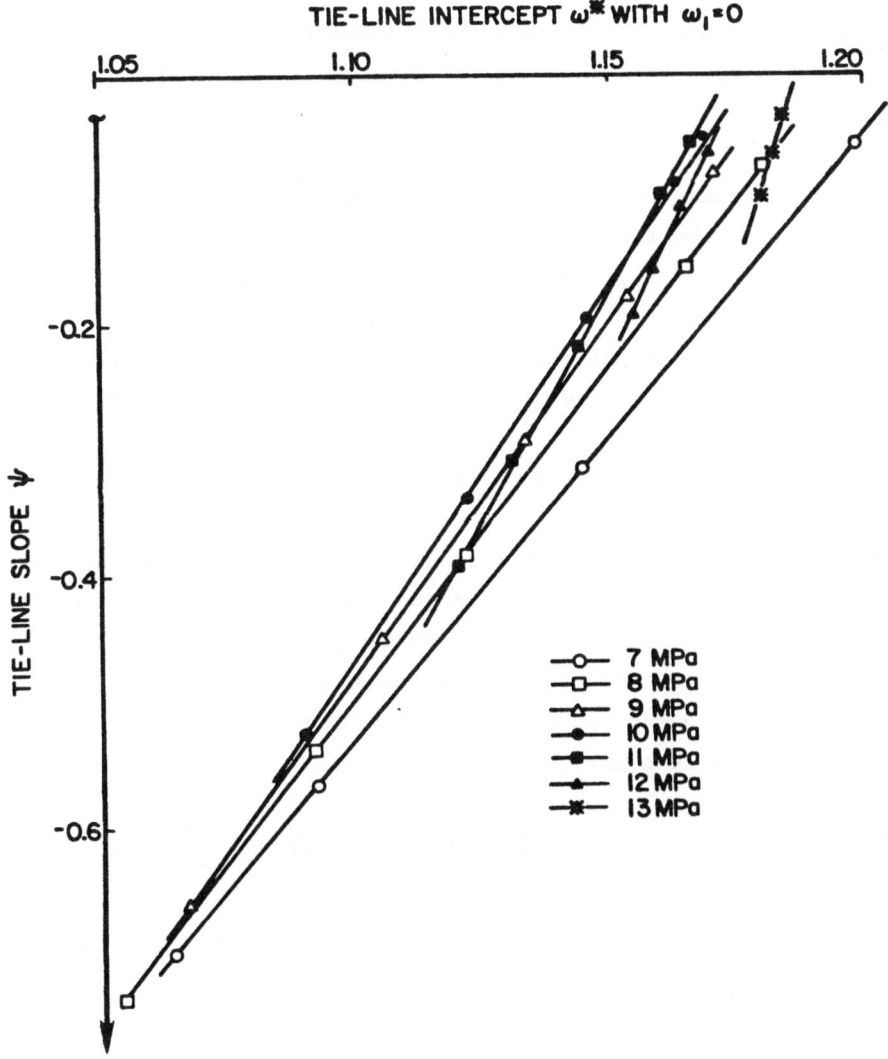

Figure 2-12. Approximate straight-line relationships
between tie-line slopes and intercepts at
various pressures for the ternary mixture
CO_2 + n-butane + n-decane at 344.26K.

approximation in this case gives a maximum relative error $\|\underset{\sim}{\psi} - \overset{\wedge}{\underset{\sim}{\psi}}\|_2 / \|\underset{\sim}{\psi}\|_2$ $\simeq 0.013$. Figure 2-13 displays α and β as functions of pressure.

These representations furnish the following procedure for calculating vapor-liquid equilibria for a point (ω_1, ω_2, p) lying in the two-phase region:

(i)   Compute the values $\alpha(p)$, $\beta(p)$ by linear interpolation

(ii)   Find the tie-line intercept $\omega*$ by solving the equation $\omega_2 = (\alpha\omega* + \beta)(\omega_1 - \omega*)$ requiring that the tie line pass through (ω_1, ω_2).

(iii) Compute the tie-line slopes from equation (2.3-5).

(iv) Find the intersections of the tie line with the Maxwell set.

Steps (ii) and (iv) deserve some comment.

The equation for $\omega*$ in step (ii) is quadratic and hence may have two distinct roots. The choice between roots will generally be clear from the construction of the tie lines. In the present case the data force $\alpha > 0$, so the slope of the tie line increases as $\omega*$ decreases. This implies that the quadratic equation may hold for some value of $\omega*$ less than the physically correct value, so we choose the largest root.

Step (iv) calls for the solution of two pairs of equations, each pair having the form

$$\omega_2^{\alpha} = \overset{\wedge}{\psi}(p)(\omega_1^{\alpha} - \omega*)$$

$$(2.3\text{-}6a)$$

$$\hat{p}^{sat}(\omega_1^{\alpha}, \omega_2^{\alpha}) = p$$

$$(2.3\text{-}6b)$$

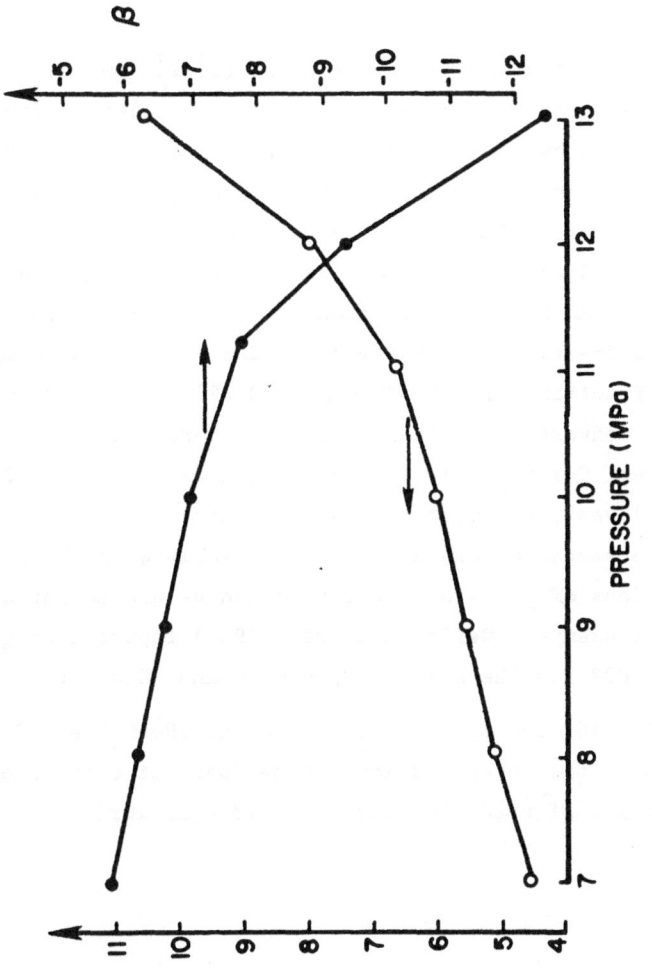

Figure 2-13. Variation of the tie-line parameters α and β with pressure for the ternary mixture CO_2 + n-butane + n-methane at 344.26K.

One simple way to solve these equations numerically is to use the method of bisection (Dahlquist et al., 1974, Section 6.2), choosing starting guesses $(\omega_1^\alpha, \omega_2^\alpha)^0$ that lie above the p^{sat} dome for the given pressure. When the bisection method has reduced the search interval to a single grid element of Ω_2^+, switching to the Newton-Raphson method gives the exact answer in one additional step since $\hat{p}^{sat}$ has uniform slope there.

Because the individual phase compositions $(\omega_1^\alpha, \omega_2^\alpha)$ in this scheme lie on an approximation to the Maxwell set, their values differ slightly from those predicted by the methods of Section 2.2. These differences lead to differences in the phase mole fractions Y_α. A comparison of results of the interpolation scheme to those of standard equation-of-state methods for 25 arbitrary points in the two-phase region of the model system shows an average root-mean-square difference in predictions for species mole fractions of $RMS(\Delta\omega_i^\alpha) \approx 0.003$. A similar comparison for phase mole fractions yields $RMS(\Delta Y_V) \approx 0.025$. These errors are comparable with convergence criteria for the iterative solution to the transport problem; Coats(1980), for example, uses $|\Delta\omega| = 0.002$. What is perhaps more relevant, the differences in predictions between the interpolation and the standard method compare favorably with the differences between predictions of the standard method and values measured in the laboratory. For example, Oellrich et al. (1981) report average deviations $|\Delta\omega_i^V| = 0.009$ for the system CO_2 + n-butane (Olds et al., 1949) and $|\Delta\omega_i^V| = 0.006$ for the system CO_2 + n-decane (Reamer and Sage, 1963). Therefore, the interpolation scheme does not significantly degrade the quality of predicted vapor-liquid equilibria.

Efficiency of the scheme.

There seems to be no wholly adequate method for comparing the
efficiency of the interpolation scheme to that of the standard approach.
However, several simple measures indicate that the interpolation scheme
is computationally much faster. The standard equal-fugacity approach
uses arithmetic operations intensively, especially in evaluating
fugacity differences and inverting the full matrices that approximate
the Jacobians at each iteration. By contrast, the interpolation scheme
requires few arithmetic operations per iteration. Thus while the bisection algorithm of the interpolation scheme has a lower convergence rate
(roughly one binary digit of accuracy per iteration) than the secant
method of the standard scheme, the difference in operation counts
overwhelms the latter's advantage in this regard.

We can try to compare the two methods on the basis of CPU time
required for vapor-liquid equilibrium calculations. In addition to
obvious questions concerning the efficiency of the author's coding
techniques, this comparison faces the further difficulty that the CPU
time required to compute a vapor-liquid equilibrium using the interpolation scheme is very small -- usually a few thousandths of a second on
the IBM 3081. This time is comparable to the time spent in calling the
CPU clock. Moreover, starting guesses for the standard method must be
fairly close to the unknown solution, while the interpolation method
starts with sure initial guesses outside the two-phase region. Still, a
sample of 25 vapor-liquid equilibria, using the final answers from the
interpolation routine as starting guesses for the standard algorithm,
gave ratios of runtime for the latter to that for the former ranging
from 15.8 to 83.4 and averaging 36.0.

These comparisons take no account of convergence failures in the standard method, which do not afflict the interpolation scheme. Perhaps the greatest improvement offered by Maxwell-set interpolation is its sure convergence.

CHAPTER THREE
SOLVING TRANSPORT EQUATIONS BY COLLOCATION

The transport equations developed in Chapter One are partial differential equations whose solutions vary in space and time. The petroleum
industry has traditionally relied on the method of finite differences to
produce discrete analogs of these equations for numerical solution.
However, some investigators have proposed finite-element Galerkin
schemes, citing the possibility of greater accuracy at comparable cost
to finite-difference methods. This chapter examines a third option,
finite-element collocation. This technique is closely related to
Galerkin methods but offers significant savings in computation. In
particular, the chapter presents a new technique, called upstream collocation, that is suitable for use in convection-dominated problems.

Numerical solution schemes for partial differential equations arising
in physics raise the thorny issue of verification. Actually there are
at least two separate problems here. First, does the discrete scheme
generate answers that are good approximations to the exact solution of
the continuous problem? Second, does any class of realizations of the
discrete scheme serve as a veracious model for an identifiable class of
observed physical events? This second problem is a discipline in
itself, and we shall not try to answer formally whether any numerical
approximation to the transport equations of Chapter One faithfully
models experiments. We shall be concerned instead with grounds for
believing that the proposed collocation scheme gives approximate
solutions to the transport equations, presuming that these equations are
adequate models.

Even within this limited scope difficulties arise. When the problem to be solved has a known exact solution, we can compare it with the approximate solution to judge the latter's acceptability. For the fully compositional equations of miscible gas flooding, however, we do not know an exact solution, and such a comparison is impossible. Nevertheless, it is perhaps reasonable to infer that a scheme gives good approximations in a complicated problem if it gives good approximations in mathematically related but analytically more tractable ones. We shall examine three such problems in this chapter, treating the fully compositional case in Chapter Four.

3.1 Orthogonal collocation on finite elements.

Orthogonal collocation is a fairly old method, due apparently to Lanczos (1938). Its adaptation to finite elements, however, is more recent (Russell and Shampine, 1972; de Boor and Swartz, 1973). There are several ways to introduce the technique; let us review it as a method of weighted residuals on a linear equation, drawing connections with the variational form of the problem.

Description of the method.

Consider a partial differential equation of the form

$$\partial_t u + E\, u = 0 \qquad \text{on } \Omega \times \Theta$$

$$(3.1\text{-}1)$$

where $\Omega = [0, x_{max}]$ is the spatial domain of the problem and $\Theta = [0, t_{max}]$ is its temporal domain. For the purpose of illustration, let E stand for a linear, second-order elliptic operator:

$$E = - \partial_x [\alpha_1(x) \ \partial_x] + \alpha_2(x) \ \partial_x$$

$$(3.1-2)$$

where α_1, $\alpha_2 \in C^\infty(\Omega)$. A typical initial boundary-value problem for equation (3.1-1) is to find a function u satisfying (3.1-1) along with auxiliary data of the form

$$u(x,0) = u_0(x), \quad x \in (0, x_{max})$$

$$(3.1-3a)$$

$$\beta_1 \ u(0,t) + \beta_2 \ \partial_x u(0,t) = \beta_3$$

$$(3.1-3b)$$

$$t \in \Theta$$

$$\beta_4 \ u(x_{max}, t) + \beta_5 \ \partial_x u(x_{max}, t) = \beta_6$$

$$(3.1-3c)$$

Taken literally, this problem requires that $u(\bullet, t) \in C^2(\Omega)$ for all $t \in \Theta$ and $u(x, \bullet) \in C^1(\Theta)$ for all $x \in \Omega$.

In the context of finite-element methods it is common to replace the literal interpretation of this problem by its variational version. In this form we relax the requirements on $u(\bullet, t)$, demanding only that it lie in the Sobolev space $H^1(\Omega)$, which contains functions $f: \Omega \to R$ such that f, $d_x f \in L^2(\Omega)$. Then u is a solution to the variational problem if (1) $u(\bullet, t) \in H^1(\Omega)$ for all $t \in \Theta$, (2) $u(x, \bullet) \in C^1(\Theta)$ for all $x \in \Omega$, (3) u satisfies the auxiliary conditions (3.1-3), and (4)

$$\int_\Omega (\partial_t u) \ v \ dx \ + \ \int_\Omega (\alpha_1 \ \partial_x u + \alpha_2 \ u) \ \partial_x v \ dx = 0$$

$$(3.1-4)$$

for any "test function" $v \in H^1(\Omega)$ that vanishes on the boundary $\partial \Omega$ (Oden and Reddy, 1976, Section 9.2).

Solving such a problem numerically requires analogs of the continuous operators that are discrete in both space and time. The appropriate spatial discretization in this case is a finite-element representation. To construct one, let $\Delta_M: 0 = \bar{x}_1 < \ldots < \bar{x}_M = x_{max}$ be a uniform partition of Ω with mesh $\Delta x = \bar{x}_\ell - \bar{x}_{\ell-1}$, and denote each element $[\bar{x}_\ell, \bar{x}_{\ell+1}] = \Omega_\ell$. Associated with the partition Δ_M is the following finite-dimensional subspace of $H^1(\Omega)$:

$$H_3(\Delta_M) = \{v \in H^1(\Omega) \mid v \text{ is a cubic polynomial over}$$
$$\text{each } \Omega_\ell, \; \ell = 1, \ldots, M\}$$

$$(3.1-5)$$

This space is the span of the Hermite cubic interpolation basis $\{H_{0,\ell}(x), H_{1,\ell}(x)\}_{\ell=1}^{M}$, whose elements are the piecewise polynomials

$$H_{0,\ell}(x) = \begin{cases} (x - \bar{x}_{\ell-1})^2 [2(\bar{x}_\ell - x) + \Delta x]/\Delta x^3, & x \in [\bar{x}_{\ell-1}, \bar{x}_\ell] \\ (\bar{x}_{\ell+1} - x)^2 [3\Delta x - 2(\bar{x}_{\ell+1} - x)]/\Delta x^3, & x \in [\bar{x}_\ell, \bar{x}_{\ell+1}] \\ 0, & x \notin [\bar{x}_{\ell-1}, \bar{x}_{\ell+1}] \end{cases}$$

$$(3.1-6a)$$

$$H_{1,\ell}(x) = \begin{cases} (x - \bar{x}_{\ell-1})^2 (x - \bar{x}_\ell)/\Delta x^2, & x \in [\bar{x}_{\ell-1}, \bar{x}_\ell] \\ (x - \bar{x}_{\ell+1})^2 (x - \bar{x}_\ell)/\Delta x^2, & x \in [\bar{x}_\ell, \bar{x}_{\ell+1}] \\ 0, & x \notin [\bar{x}_{\ell-1}, \bar{x}_{\ell+1}] \end{cases}$$

$$(3.1-6b)$$

Thus every function $v \in H_3(\Delta_M)$ is a finite linear combination

$$v(x) = \sum_{\ell=1}^{M} [v_\ell \, H_{0,\ell}(x) + v'_\ell \, H_{1,\ell}(x)]$$

$$(3.1-7)$$

where the unique coefficients v_ℓ and v'_ℓ are the nodal values of v and its gradient $d_x v$, respectively, at the node $\bar{x}_\ell$. Moreover, as $\Delta x \to 0$, $H_3(\Delta_M)$ is dense as a subspace of $H^1(\Omega)$ in the Sobolev norm $\|\bullet\|_{2,1}$ defined by $\|f\|_{2,1}^2 = \|f\|_{L^2(\Omega)}^2 + \|d_x f\|_{L^2(\Omega)}^2$. This fact guarantees that finite-element approximations in $H_3(\Delta_M)$ will be consistent (Oden and Reddy, 1976, Section 8.3). Finally, there is an interpolation operator associated with $H_3(\Delta_M)$, namely, the projection I_M: $C^1(\Omega) \to H_3(\Delta_M)$ mapping functions $f \in C^1(\Omega)$ onto their continuously differentiable Hermite interpolates:

$$(I_M f)(x) = \sum_{\ell=1}^{M} [f(\bar{x}_\ell) \, H_{0,\ell}(x) + d_x f(\bar{x}_\ell) \, H_{1,\ell}(x)]$$

$$(3.1-8)$$

In terms of this finite-element formalism, the spatially discrete analog to the initial boundary-value problem (3.1-1) and (3.1-3) is to find a trial function $\hat{u}(\bullet,t) \in H_3(\Delta_M)$ that best approximates u in the sense of weighted residuals, that is,

$$\int_\Omega [\partial_t \hat{u}(x,t) + E \, \hat{u}(x,t)] \, \delta_k(x) \, dx = 0, \qquad k = 1,\ldots 2M-2$$

$$(3.1-9)$$

$$\hat{u}(x,0) = I_M \, u_0(x), \qquad x \in (0,x_{max}),$$

$$(3.1-10)$$

along with the boundary data (3.1-3a) and (3.1-3b), for a collection of weight functions $\delta_k(x)$. In orthogonal collocation the weight functions are the Dirac distributions $\delta_k(x) = \delta(x - x_k)$, where the collocation points x_k are the $K_c = 2(M - 1)$ Gauss points $\bar{x}_\ell + \frac{1}{2} \Delta x \pm \Delta x/\sqrt{3}$, $\ell = 1,\ldots,M-1$. This method leads to a set of ordinary differential equations for the evolution of the approximating function u:

$$d_t \hat{u}(x_k,t) + E \, \hat{u}(x_k,t) = 0, \quad k = 1,\ldots,K_c$$

(3.1-11)

with the initial condition (3.1-10). As we shall review below, choosing the Gauss points as collocation points leads to the highest possible order of spatial accuracy.

It is worth observing briefly that the method of weighted residuals is, in a sense, a generalized discrete form of the variational problem (3.1-4) (Strang and Fix, 1973, Section 2.3). In the case where the weight functions are basis functions we recover Galerkin's method, which is a direct finite-dimensional analog of (3.1-4). The connection between this particular case and weighted-residual techniques involving other weight functions depends in general on the choice of weight functions. For collocation, however, there is a well defined correspondence with Galerkin's method and hence with the variational formulation. Appendix C explains this correspondence.

To discretize equations (3.1-11) in time let us use one of three finite-difference schemes. Call $\hat{u}^n(x) = \hat{u}(n \, \Delta t,x)$, where Δt is a time increment. The explicit Euler scheme for (3.1-11) is

$$\hat{u}^{n+1}(x_k) - \hat{u}^n(x_k) + \Delta t \, E \, \hat{u}^n(x_k) = 0, \quad k = 1,\ldots,K_c,$$

(3.1-12)

the implicit Euler scheme is

$$\hat{u}^{n+1}(x_k) - \hat{u}^n(x_k) + \Delta t \, E \, \hat{u}^{n+1}(x_k) = 0, \quad k = 1,\ldots,K_c, \tag{3.1-13}$$

and the Crank-Nicolson scheme is

$$\hat{u}^{n+1}(x_k) - \hat{u}^n(x_k)$$

$$+ \quad \Delta t \, [E \, \hat{u}^{n+1}(x_k) + E \, \hat{u}^n(x_k)] = 0, \quad k = 1,\ldots,K_c. \tag{3.1-14}$$

The two Euler schemes have temporal truncation error $O(\Delta t)$, while the Crank-Nicolson scheme has truncation error $O(\Delta t^2)$.

Performance of the scheme.

Like other finite-element methods, discrete-time schemes such as (3.1-12) through (3.1-14) convert a continuous problem to a set of algebraic equations at each time level. If the original problem is linear, or if we linearize a nonlinear problem and iterate, the discrete problem reduces to the inversion of matrices that are sparse owing to the limited support of the basis functions. For the Hermite cubic basis in one dimension the matrices have a bandwidth of five as drawn in Figure 3-1. This sparsity is attractive since it allows the use of efficient inversion algorithms.

Orthogonal collocation on finite elements offers the further advantage of high-order spatial accuracy. Douglas and Dupont (1973), for example, investigate the semidiscrete method (3.1-11) for quasilinear parabolic equations in one spatial dimension. Their analysis rests on the fact that finite-element collocation at the Gauss points is algebraically equivalent to an approximate Galerkin method in which the

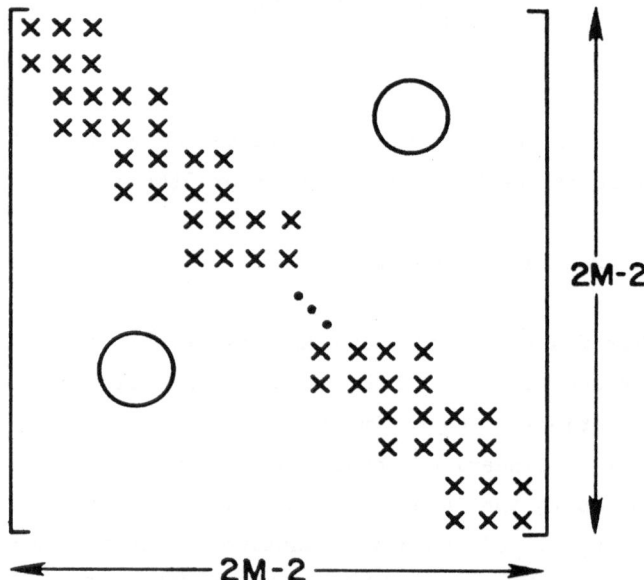

Figure 3-1. Structure of the coefficient matrix for finite-element collocation on Hermite cubics in one space dimension. The symbol "x" represents a nonzero entry; M is the number of nodes.

integrals $\int_\Omega H_{i,j} H_{k,\ell}\, dx$, $\int_\Omega (d_x H_{i,j}) H_{k,\ell}\, dx$, $\int_\Omega (d_x^2 H_{i,j}) H_{k,\ell}\, dx$
contributing to the matrix elements are replaced by two-point Gauss
quadratures on each element (see Appendix C). Douglas and Dupont show
that orthogonal collocation yields approximations u to the exact
solution u that satisfy

$$\| u - \hat{u} \|_\infty = O(\Delta x^4)$$

(3.1-15)

uniformly on $\Omega \times \Theta$, provided certain smoothness conditions hold. This
error is optimal in the sense that its order is the same as that of the
interpolation errror $\| u - I_M u \|_\infty$. In particular it is the same order of
accuracy as that offered by Galerkin's method on Hermite cubics.

Indeed, the possibility of greater accuracy has attracted interest in
C^1 Galerkin techniques in oilfield problems for over a decade. Several
studies suggest that Galerkin methods on Hermite cubic spaces require
less computational cost for a given degree of accuracy than standard
finite-difference methods. Among these are papers by Cavendish et al.
(1969), who solve the elliptic equation for single-phase steady reservoir flow; Culham and Varga (1971), who treat a nonlinear gas-flow
equation; Settari et al. (1977), who solve a parabolic equation
governing miscible displacement, and Spivak et al. (1977), who investigate the coupled elliptic and parabolic equations modeling immiscible
two-phase flows in porous media.

Though offering similar advantages in accuracy, orthogonal collocation requires less computational effort than the related Galerkin
scheme. To begin with, collocation obviates the integrations needed to
compute the Galerkin element matrices. Collocation also bypasses the
formal, element-by-element assembly of the global Galerkin matrix. This

promise of greater efficiency has led to scattered applications of orthogonal collocation in engineering (see, for examples, Chawla et al., 1975; Banjia et al., 1978, and Pinder et al., 1978). Sincovec (1977) applied orthogonal collocation to several problems in petroleum engineering, confirming collocation's efficiency in two parabolic examples. However, his formulation failed to solve the hyperbolic Buckley-Leverett problem. This result may have discouraged broader application of collocation to oilfield problems.

3.2. The convection-dispersion equation.

The convection-dispersion equation is one of the simplest linear equations featuring the time dependence and convective dominance characteristic of many porous-media flows. Discrete methods with high-order spatial accuracy often yield qualitatively flawed solutions to such problems when convection is strong. As we shall see, orthogonal collocation is no exception to this rule.

Physical setting.

The convection-dispersion equation is a special case of the species transport equation (1.5-3) when (1) only one fluid phase is present; (2) the fluid velocity is known and constant; (3) the effects of density changes and gravity are negligible, and (4) dispersion is Fickian with a constant dispersion coefficient. Under these assumptions, the concentration ω_i of a species dissolved in the fluid obeys the parabolic equation

$$\partial_t \omega_i = D_i \, \partial_x^2 \omega_i - v \, \partial_x \omega_i$$

$$(3.2-1)$$

over some space-time domain $[0, x_{max}] \times [0, t_{max}]$. This equation reduces to a dimensionless form

$$\partial_\tau \omega_i = (Pe^{-1}) \, \partial_x^2 \omega_i - \partial_x \omega_i$$

$$(3.2-2)$$

where

$$\chi = x / x_{max}$$

$$(3.2-3a)$$

$$\tau = v t / t_{max}$$

$$(3.2-3b)$$

and $Pe = v x_{max} / D_i$ is the Peclet number. In this form it is apparent that the convection-dispersion equation is a singular perturbation of a first-order hyperbolic equation corresponding to the limit $Pe \rightarrow \infty$. For many applications Pe is quite large, and numerical solutions exhibit the types of errors associated with the hyperbolic limit. In such convection-dominated problems large gradients in the exact solution tend to persist, and a spatial discretization too coarse to resolve steep portions of the initial data will fail to propagate them correctly (Gray and Pinder, 1976).

Solution using orthogonal collocation.

To illustrate the difficulty with steep initial data, let us apply orthogonal collocation to equation (3.2-1) with the auxiliary data

$$\omega_i(x,0) = 0, \quad x \in (0,x_{max})$$

$$(3.2-4a)$$

$$\omega_i(0,t) = 1, \quad t \geq 0$$

$$(3.2-4b)$$

$$\partial_x \omega_i(x_{max},t) = 0, \quad t \geq 0$$

$$(3.2-4c)$$

The Hermite cubic trial function for this problem has the form

$$\hat{\omega}(x,t) = \sum_{\ell=1}^{M} [W_\ell(t) H_{0,\ell}(x) + W'_\ell H_{1,\ell}(x)]$$

$$(3.2-5)$$

where the coefficients $W_\ell(t)$, $W'_\ell(t)$ represent the time-dependent values of ω, $\partial_x\omega$, respectively, at the node $\bar{x}_\ell$. Substituting $\hat{\omega}$ into (3.2-1) and collocating gives a set of ordinary differential equations for $\{W_\ell, W'_\ell\}_{\ell=1}^{M}$:

$$\sum_{\ell=1}^{M} \{d_t W_\ell(t) \, H_{0,\ell}(x_k) + d_t W'_\ell(t) \, H_{1,\ell}(x_k)$$

$$+ v \, [W_\ell(t) \, d_x H_{0,\ell}(x_k) + W'_\ell(t) \, d_x H_{1,\ell}(x_k)]$$

$$- D_i \, [W_\ell(t) \, d_x^2 H_{0,\ell}(x_k) + W'_\ell(t) \, d_x^2 H_{1,\ell}(x_k)]\} = 0$$

$$(3.2\text{-}6)$$

$$k = 1,\ldots,2M\text{-}2$$

The boundary data imply $\hat{\omega}(0,t) = W_1(t) = 1$ and $\partial_x \hat{\omega}(x_{max},t) = W'_M = 0$, so we have 2M equations in 2M variables.

However, the initial data (3.2-4a) are not differentiable at $x = 0$, so strictly speaking the interpolate $I_M \omega_i(x,0)$ is undefined. We can circumvent this snag by assigning $W_1(t) = 1$ and $W'_1(0) = W_\ell(0) = W'_\ell(0) = 0$ for $\ell = 2,\ldots,M$, reasoning that this choice preserves monotonicity and is consistent with the true initial data as $\Delta x \to 0$. Approximating the discontinuous initial data in this way adds artificial mass $\int_\Omega \hat{\omega}(x,0) \, dx = \int_\Omega H_{0,1}(x) \, dx = O(\Delta x)$ to the solution.

Figure 3-2 shows an approximate solution to the boundary-value problem defined by (3.2-1) and (3.2-4) using orthogonal collocation for the spatial approximation and a Crank-Nicolson scheme in time. Here $v = 0.369$ m/s, $D_i = 0.000345$ m^2/s, $\Delta x = 0.25$ m, and $\Delta t = 0.5$ s. These values for v and D_i are common in the literature on the convection-dispersion equation; see for example Pinder and Gray (1977, Section 5.3). The figure also displays an approximation

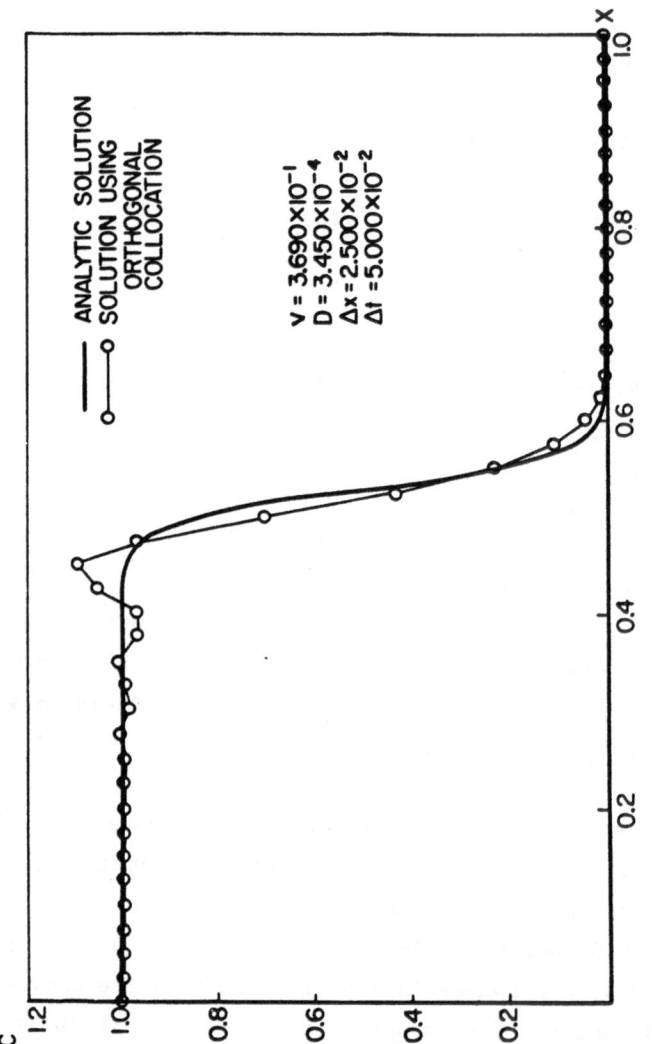

Figure 3-2. Solution to the convection-dispersion equation using orthogonal collocation.

$$\omega_i(x, t + t_0) \simeq \tfrac{1}{2} \text{ erfc } \{[x - v(t + t_0)]/\sqrt{[4D_i(t - t_0)]}\}$$

$$(3.2-7)$$

to the exact solution, where the choice of t_0 ensures that
$\int_\Omega \omega_i(x, t_0) \, dx = \int_\Omega \hat{\omega}(x, 0) \, dx$ to help check that the numerical scheme
correctly propagates the sharp front. For these parameters Pe $\simeq$ 1070, a
value indicating strongly convective flow. Upstream of the steep front
the collocation solution exhibits wiggles that contradict both physical
sense and the maximum principle for parabolic equations (John, 1981,
Section 7.1). Such wiggles are typical of high-order spatial approximations to nearly hyperbolic equations. The wiggles arise when the grid
Peclet number $\text{Pe}_{\Delta x} = v \, \Delta x / D_i$ ($\simeq$ 250 in Figure 3-2) exceeds a critical
value generally of order unity (Jensen and Finlayson, 1980).

Wiggles: pro and con.

There is some controversy regarding the desirability of suppressing
these wiggles by choice of numerical approximations. Gresho and Lee
(1979), for instance, argue against suppressing the wiggles with
low-order discretization, claiming that they should be viewed as
symptoms of overly coarse spatial grids. While there is much merit in
this argument, there are applications in which the choice between
wiggles and a sufficiently fine grid is too confining. One example in
computational physics is the gasdynamic shock, where the appropriate
grid Peclet number is comparable to the mean free path of the gas
molecules (Ames, 1977, Section 5-2). Less dramatic examples arise in
porous-media flows, where the dissipative influences of hydrodynamic
dispersion or capillarity are often genuinely small compared to convective effects. In these cases suppressing the wiggles may give acceptable answers on affordable grids, with quantifiable errors, even when
the equations to be solved are highly nonlinear. By contrast, insisting
on high-order spatial accuracy at all points of the flow field may force

a choice between exorbitantly fine grids and instability, as the wiggles
associated with coarse grids may cause associated thermodynamic calculations to fail.

Furthermore, as Section 3.4 discusses, globally high-order methods
may converge to solutions that are altogether incorrect. Such anomalies
arise in nonlinear convection-dominated problems in which the dissipative effects, while numerically small, exert physically important influences near sharp fronts, as reviewed in Section 1.5. In these cases the
failure to converge is arguably a more serious shortcoming than the
propagation of wiggles. Therefore, while several investigators have
advanced schemes to correct the wiggles without dissipation (see, for
example, Van Genuchten and Gray, 1978; Chaudhari, 1971; Boris and Book,
1976), we shall be interested specifically in a scheme that adds dissipation in the form of a low-order spatial error.

What is wanted in many cases is a method that has locally low-order
spatial accuracy. In other words, we seek a scheme whose low-order
error terms, while present globally, have small coefficients except in
the spatial zones where they are needed. Such schemes retain many of
the advantages of high-order schemes while adding dissipation near sharp
fronts. For the convection-dispersion equation, for example, a lowest-
order error term proportional to $v \, \Delta x \, \partial_x^2 \omega$ would be appropriate. In
finite-difference theory locally low-order schemes are available through
the use of upstream-weighted differences. We shall discuss an analogous
finite-element method, upstream collocation.

3.3. Upstream collocation.

Consider the following modification to equation (3.2-6):

$$\sum_{\ell=1}^{M} \{d_t W_\ell(t) \, H_{0,\ell}(x_k) + d_t W'_\ell(t) \, H_{1,\ell}(x_k)$$

$$+ v \, [W_\ell(t) \, d_x H_{0,\ell}(x_k^*) + W'_\ell(t) \, d_x H_{1,\ell}(x_k^*)]$$

$$- D_i \, [W_\ell(t) \, d_x^2 H_{0,\ell}(x_k) + W'_\ell(t) \, d_x^2 H_{1,\ell}(x_k)]\} = 0$$

$$k = 1,\ldots,2M-2$$

$$(3.3-1)$$

Here $x_k^* = x_k - \zeta \, \Delta x$, with $\zeta > 0$ chosen so that x_k^* lies in the same interval Ω_ℓ of the partition Δ_M as x_k. The difference between equations (3.3-1) and (3.2-6) is that in the latter the collocation points for the convective terms lie upstream of the Gauss points. The heuristic for choosing the x_k^* in this way is an analogy with finite differences, where upstream-weighted differences offer one approach to suppressing wiggles at the expense of smearing in highly convective problems (Peaceman, 1977, Chapter 4).

Figure 3-3 shows solutions of the same convection-dispersion problem plotted in Figure 3-2 except for the upstream choices of x_k^*. The figure specifies the collocation points in terms of the local space coordinate ξ, defined on any element $[\bar{x}_\ell, \bar{x}_{\ell+1}]$ as

$$\xi(x) = [2(x - \bar{x}_\ell)/\Delta x] - 1$$

$$(3.3-2)$$

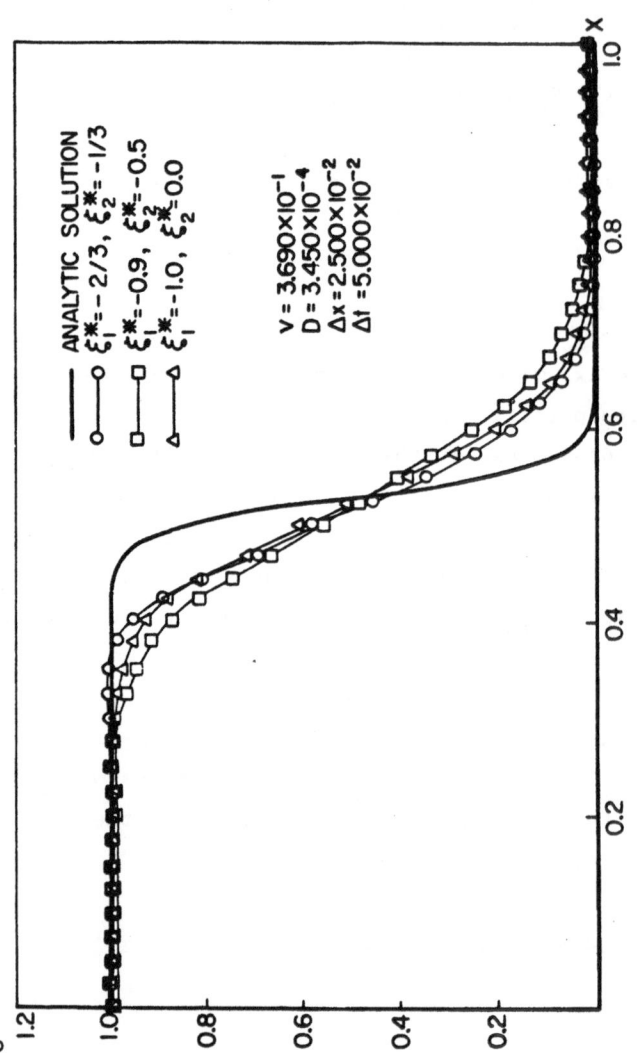

Figure 3-3. Solutions to the convection-dispersion equation using several choices of upstream collocation points.

so that $\xi: [\bar{x}_\ell, \bar{x}_{\ell+1}] \to [-1,1]$. In local terms, the orthogonal collocation points are $\xi_k = \pm 1/\sqrt{3}$ in each element, and the upstream points are
$\xi_k{}^* = \xi(x_k{}^*) < \xi_k$. The solutions in Figure 3-3 confirm the heuristic
for upstream collocation: by emphasizing upstream values in the convective terms we have suppressed the wiggles generated by orthogonal collocation, the cost being a diffusion-like smearing of the steep front.

This smearing reflects the nature of the error induced by upstream
collocation. By Taylor's theorem,

$$d_x H_{j,\ell}(x_k{}^*) = d_x H_{j,\ell}(x_k) - \zeta \, \Delta x \, d_x^2 H_{j,\ell}(x_k)$$

$$+ \tfrac{1}{2} \zeta^2 \Delta x^2 d_x^3 H_{j,\ell}(x_k), \qquad j = 0,1$$

$$(3.3-3)$$

exactly, since each basis function $H_{j,\ell}$ is cubic. Therefore upstream
collocation for the convection-dispersion equation is equivalent to the
following scheme:

$$\sum_{\ell=1}^{M} \{ d_t W_\ell(t) \, H_{0,\ell}(x_k) + d_t W'_\ell(t) \, H_{1,\ell}(x_k)$$

$$+ v \, [W_\ell(t) \, d_x H_{0,\ell}(x_k) + W'_\ell(t) \, d_x H_{1,\ell}(x_k)]$$

$$- (D_i + \zeta v \Delta x) \, [W_\ell(t) \, d_x^2 H_{0,\ell}(x_k) + W'_\ell(t) \, d_x^2 H_{1,\ell}(x_k)]\}$$

$$= O(\Delta x^2), \qquad k = 1, \ldots, 2M-2$$

$$(3.3-4)$$

In other words, upstream collocation augments the physical dissipation
by an amount proportional to Δx, which is precisely the effect of
upstream-weighted differencing (Lantz, 1971). The author has verified

this error estimate through a numerical example, finding good agreement between the value of $D_i + \zeta\, v\, \Delta x$ and the observed level of dissipation in a numerical solution (Allen, 1983).

The error introduced in upstream collocation is closely analogous to that of some Galerkin schemes. Most notable among these are the upwind quadrature techniques developed by Hughes (1978), Hughes and Brooks (1979), and Dalen (1979), who approximate the Galerkin integrals arising from convective terms by finite sums emphasizing upstream values of the integrands. There is a well defined correspondence between these quadrature methods and upstream collocation through the algebraic equivalence with approximate Galerkin schemes exhibited in Appendix C. Less exact analogies stand between upstream collocation and finite-element techniques based on the use of asymmetric basis functions for convective terms. Huyakorn (1977), Heinrich and Zienkiewicz (1977), and Christie and Mitchell (1978) have proposed such schemes for C^0 and C^1 Galerkin approximations, and Shapiro and Pinder (1981, 1982) have examined a related approach using collocation. These asymmetric basis methods have similar effects in solving convection-dominated flows, but their algebraic relationship to upstream collocation is not straightforward.

3.4. The Buckley-Leverett problem.

While our discussion of collocation has so far concentrated on linear
parabolic problems, the problems in miscible gas flood engineering are
more frequently nonlinear. Moreover, as we have discussed, problems of
practical interest in this field are often highly convective singular
perturbations of hyperbolic problems. The Buckley-Leverett problem is a
simple, analytically solvable paradigm of a nonlinear hyperbolic conservation law arising in porous-media physics. While solving the problem
numerically is of little practical interest per se, it is a reasonable
test of whether a candidate numerical method is appropriate for more
complicated reservoir flows. We shall demonstrate that upstream collocation corrects a tendency of orthogonal collocation to converge to
incorrect solutions in this problem. The key to the improvement offered
by upstream collocation is its lowest-order error term, which acts in a
manner paralleling the artificial viscosity method of finite differences
(Allen and Pinder, 1982).

Physics of the problem.

The Buckley-Leverett saturation equation governs the flow of two
immiscible, incompressible fluids, say a vapor and a liquid, in a
homogeneous porous medium. The derivation given in Section 1.5 can be
extended to include the effects of gradients in capillary pressure p_{CVL},
the result being the hyperbolic equation (1.5-10) for the vapor saturation S_V augmented by a second-order term:

$$\partial_t S_V + \partial_x \{Q\, f_V\, [1 + \Lambda_L\, (dp_{CVL}/dS_V)\, \partial_x S_V]\} = 0$$

$$(3.4-1)$$

where, as in equation (1.5-11), $f_V = \Lambda_V/(\Lambda_V + \Lambda_L) = f_V(S_V)$ is the
fractional flow function, Λ_α is the mobility of phase α, and Q is the

constant total flow rate divided by the porosity of the rock. The first-order form arises when the capillary gradient $(dp_{CVL}/dS_V)\ \partial_x S_V$ contributes negligibly to the fluid motions, a situation occurring in practice when applied pressure gradients give rise to high Darcy velocities. In this case we recover the hyperbolic conservation law

$$\partial_t S_V + \partial_x (Q f_V) = 0$$

$$(3.4-2)$$

As Section 1.5 reviews, the fact that $f_V(S_V)$ is not convex over its support leads to the formation of discontinuities in $S_V(x,t)$ given steep initial data. Therefore we must accept weak solutions to (3.4-2), guaranteeing uniqueness only by imposing in addition some form of the shock condition. The form that we shall discuss is the demand that the solution to (3.4-2) be the limit as $\eta \to 0$ of solutions to the parabolic extension

$$\partial_t S_V + \partial_x (Q f_V) = \partial_x (\eta\ \partial_x S_V)$$

$$(3.4-3)$$

with $\eta > 0$. Viewed from the standpoint of mechanics, the term on the right side of (3.4-3) is analogous to a capillary influence of the type appearing in the full equation (3.2-1), except for spatial variations in the coefficient $f_V \Lambda_L (dp_{CVL}/dS_V)$. Heuristically, the shock condition "restores" the problem-closing effects of capillarity in models of physical flows where its action, while profound, is limited to a practically infinitesimal zone where $\partial_x S_V$ is virtually infinite.

Discretization.

Consider the Cauchy problem for equation (3.4-2) on $\Omega \times \Theta = [0,\infty) \times [0,\infty)$ with the initial data

$$S(0,t) = 1 - S_{LR}, \quad t \geq 0$$

$$(3.4\text{-}4a)$$

$$S(x,0) = S_{VR}, \quad x > 0$$

$$(3.4\text{-}4b)$$

Here, as in Chapter One, S_{VR} and S_{LR} are the residual vapor and liquid saturations, respectively. Let us assume an S-shaped fractional flow function f_V determined by the mobilities

$$A_V(S_V) = (S_V - S_{VR})^2$$

$$(3.4\text{-}5a)$$

$$A_L(S_V) = (1 - S_V - S_{LR})^2$$

$$(3.4\text{-}5b)$$

with $S_{VR} = 0.16$, $S_{LR} = 0.20$, and a flow rate $Q = 2.134 \times 10^{-4}$ m/s.

Discretizing (3.4-2) in time gives

$$\Delta_t S_V + Q \, \Delta t \, \partial_x f_V = 0$$

$$(3.4\text{-}6)$$

where $\Delta_t S_V(x,t) = S_V(x,t+\Delta t) - S_V(x,t)$. Then, representing $\Delta_t S_V$ by a Hermite trial function, we have

$$\Delta_t S_V(x,t) \simeq \Delta_t \hat{S}(x,t) = \sum_{\ell=1}^{M} [\psi_\ell(t) \, H_{0,\ell}(x) + \psi'_\ell(t) \, H_{1,\ell}(x)]$$

$$(3.4\text{-}7)$$

This representation furnishes a Hermite approximation to the saturation,

$$S_V(x,t) \simeq \hat{S}(x,t) = \sum_{\ell=1}^{M} [\Psi_\ell(t) \, H_{0,\ell}(x) + \Psi'_\ell(t) \, H_{1,\ell}(x)]$$

$$(3.4\text{-}8)$$

via the updating rules

$$\Psi_\ell(t + \Delta t) = \Psi_\ell(t) + \psi_\ell(t + \Delta t)$$

$$\text{(3.4-9a)}$$

$$\Psi'_\ell(t + \Delta t) = \Psi'_\ell(t) + \psi'_\ell(t + \Delta t)$$

$$\text{(3.4-9b)}$$

For the flux term in (3.4-6) let us use an approximation of the form

$$\hat{f}(x,t+\Delta t) = \sum_{\ell=1}^{M} [f_V(\Psi_\ell(t)) H_{0,\ell}(x) + f_V'(\Psi_\ell(t)) \Psi'_\ell(t) H_{1,\ell}(x)]$$

$$+ [f_V' \Delta_t \hat{S_V}](x,t)$$

$$\text{(3.4-10)}$$

where the last term is a linear projection to time $t + \Delta t$ represented by the product expansion

$$[f_V' \Delta_t \hat{S_V}](x,t) = \sum_{\ell=1}^{M} \{[f_V'(\Psi_\ell(t)) \psi_\ell(t+\Delta t)] H_{0,\ell}(x)$$

$$+ [f_V'(\Psi_\ell(t)) \psi'_\ell(t+\Delta t) + f_V''(\Psi_\ell(t)) \Psi'_\ell(t) \psi_\ell(t+\Delta t)] H_{1,\ell}(x)\}$$

$$\text{(3.4-11)}$$

The Hermite gradient coefficients in (3.4-11) are x-derivatives developed formally using the chain rule. This linearized implicit treatment is equivalent to one Newton-Raphson iteration per time step. Substituting the Hermite representations (3.4-7), (3.4-8), and (3.4-10) and collocating reduces equation (3.4-6) to the following set of algebraic equations:

$$Q \, \Delta t \, \partial_x f(x_k^*, \, t + \Delta t) + \Delta_t S(x_k, \, t + \Delta t) = 0,$$

$$(3.4\text{-}12)$$

$$k = 1, \ldots, 2M-2$$

where x_k^* signifies an upstream collocation point. These equations, together with the Cauchy data, must give a closed linear system for the unknowns $\{\psi_\ell, \psi'_\ell\}_{\ell=1}^M$ at each time step.

The initial data (3.4-4) pose two difficulties in this regard. One of these is similar to that encountered in interpolating the initial conditions (3.2-4a) for the convection-dispersion equation: $S_V(x,0) \notin C^1(\Omega)$, so its interpolate $I_M S_V$ is not strictly defined. As in the convection-dispersion problem, we can impose consistent initial conditions in $H_3(\Delta_M)$ by prescribing $\Psi_1(t) = 1 - S_{LR}$, $\Psi'_1(t) = \Psi_\ell(0) = \Psi'_\ell(0) = S_{VR}$ for $\ell = 2, \ldots, M$. Again, this prescription adds artificial mass to the numerical solution.

The second problem with the initial data is somewhat more subtle. Digital computations require a finite number of unknowns, while (3.4-4b) specifies a semi-infinite spatial domain. The only compromise available with a uniform grid is to solve the problem on a finite spatial domain $[0, x_{max}]$ at any given time step, either increasing x_{max} as t increases or stopping the calculations before the influence of the spurious right boundary contaminates the results. There then arises the issue of boundary data at $x = x_{max}$. Any such data will necessarily be artificial for the first-order Cauchy problem, but they are indispensable in defining an algebraically closed discrete analog with 2M - 2 collocation points. Bearing in mind that the hyperbolic problem is really an extreme test of a method intended for solving parabolic equations, let us impose the artificial condition $\partial_x S_V = 0$ at $x = x_{max}$. As an alternative one might use an additional collocation point in the element abutting the endpoint $x = x_{max}$, thereby increasing the number of algebraic equations and obviating the artificial condition.

Numerical solutions.

Figure 3-4 shows a numerical solution to the Buckley-Leverett problem at t = 1500 s using orthogonal collocation ($x_k^* = x_k$) on a grid of mesh $\Delta x = 0.05$ m with $\Delta t = 5.0$ s. The figure also shows the exact solution, computed using the method of characteristics and Welge's tangent construction as described in Section 1.5. The most glaring fact in this plot is the difference between the numerical and exact predictions for the saturation shock: the numerical shock occurs upstream of the correct shock and is too strong.

Similar errors occur in spatially centered finite-difference (Aziz and Settari, 1979, Section 5.5.1) and Galerkin (Spivak et al., 1977) approximations to capillarity-free two-phase flow. These erroneous predictions reflect the failure of globally high-order spatial approximations to accommodate the shock condition needed to specify the unique, physically correct weak solution to the problem. Finite-difference models in the petroleum industry have traditionally used upstream-weighted differences to rectify this failing (Aziz and Settari, 1979, Section 5.5.1; Peaceman, 1977, Chapter 6). Galerkin models have often resorted to the explicit addition of artificial capillarity (Douglas et al., 1979; Spivak et al., 1977; Chase, 1979; Mercer and Faust, 1976), although some investigators have used various upstreaming techniques (Dalen, 1979; Chavent and Salzano, 1982).

Figure 3-5 shows approximate solutions to the same Cauchy problem using various choices of upstream collocation points x_k^*. Although these solutions vary in quality, they give much better predictions of the strength and location of the saturation shock than does orthogonal collocation. Figure 3-6 displays solutions to the Buckley-Leverett problem with the upstream collocation points $\xi_1^* = -2/3$, $\xi_2^* = 1/3$ in local element coordinates for different values of the spatial mesh Δx.

115

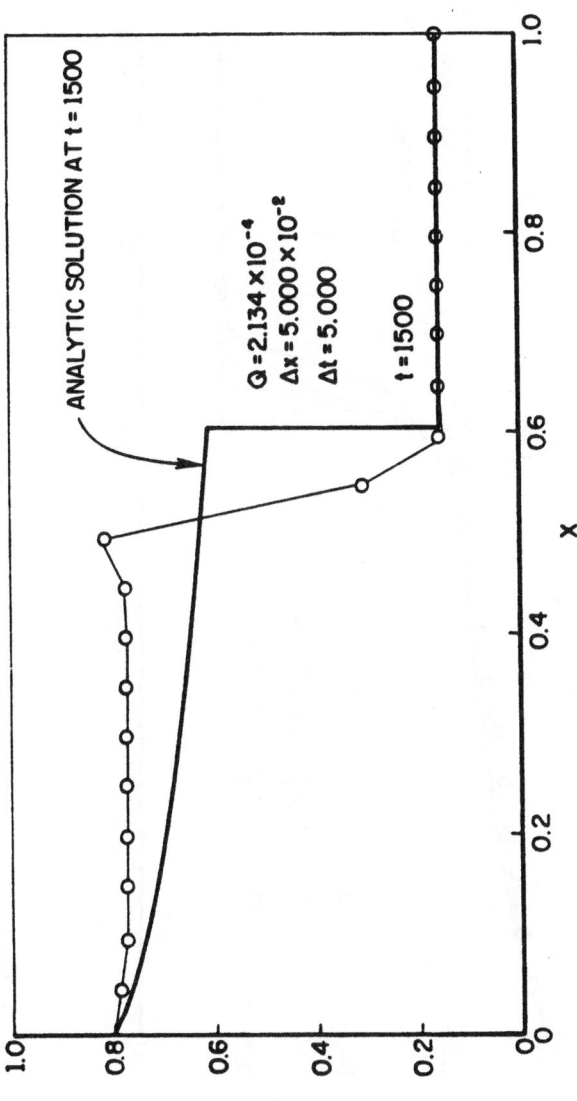

Figure 3-4. Solution to the Buckley-Leverett problem using orthogonal collocation.

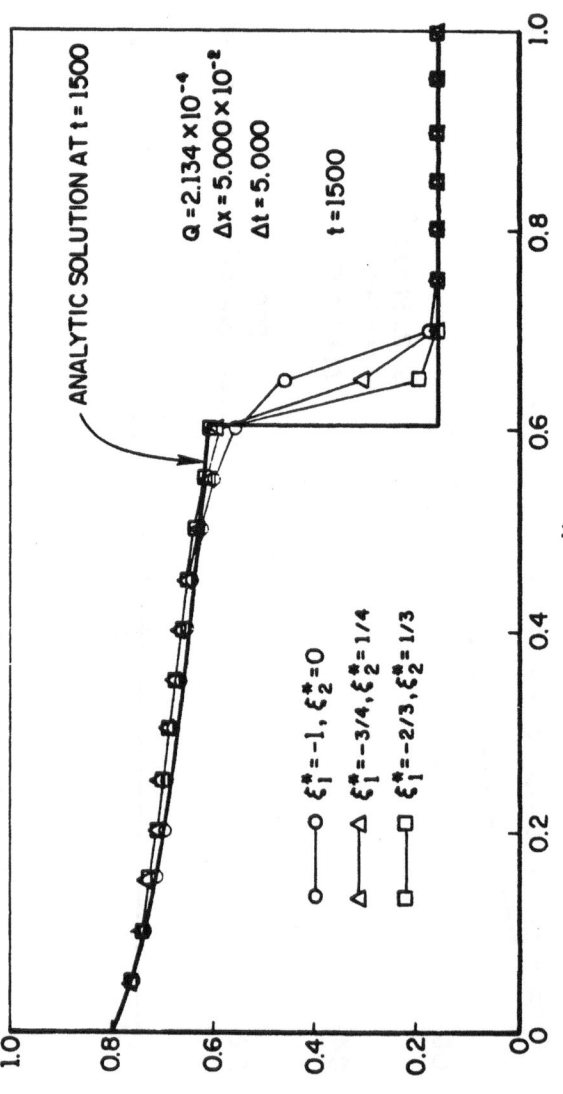

Figure 3-5. Solutions to the Buckley-Leverett problem using several choices of upstream collocation points.

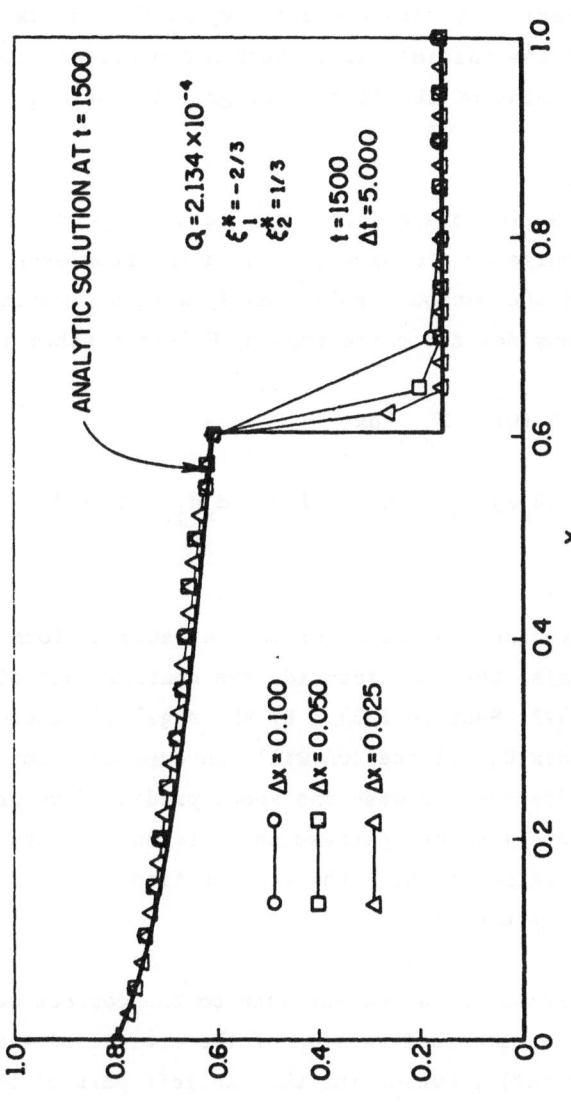

Figure 3-6. Solutions to the Buckley-Leverett problem using upstream collocation with spatial grids of varying mesh.

This plot suggests that the resolution of the correct saturation shock improves on refinement of the grid. All of these solutions consistently exhibit artificial mass manifested as a saturation "toe" downstream of the shock. To check for material balance errors we can compare this additional mass numerically with the quantity $\int_\Omega H_{0,1}(x)\, dx$ attributable to interpolation of the initial data. Such a comparison shows agreement to within half a percent of the latter integral in the worst case (Allen and Pinder, 1983).

These results deserve some discussion. We can expect finite-element collocation to converge to a weak solution to the Buckley-Leverett solution because of the method's relationship with the variational form of the problem. Consider the corresponding Galerkin scheme:

$$\int_\Omega [\partial_t \hat{S} + \partial_x(Q\hat{f})]\, H_{j,\ell}\, dx$$

$$= \int_\Omega (\partial_t \hat{S})\, H_{j,\ell}\, dx + \int_\Omega Q\, \hat{f}\, d_x H_{j,\ell}\, dx = 0$$

$$(3.4\text{-}13)$$

This is a spatially discrete analog of the variational form of equation (3.4-2), which is also the weak form for the spatial part of the problem (Strang and Fix, 1973, Section 2.3). By the algebraic correspondence described in Appendix C, collocation will converge to solutions of this weak form. The difference between the shock predicted by orthogonal collocation and that given by upstream collocation reflects the lack of uniqueness among weak solutions: the schemes find different weak solutions to the same equation.

Upstream collocation forces convergence to the correct weak solution because it imposes a numerical version of the artificial viscosity condition. Specifically, evaluating the explicit part of the flux term at upstream collocation points $x_k^* = x_k - \zeta\, \Delta x$ induces an error

$$\partial_x \hat{f}(x_k,t) - \partial_x \hat{f}(x_k^*,t) = \zeta \, \Delta x \, \partial_x^2 \hat{f}(x_k,t) + O(\Delta x^2)$$

$$= \zeta \, \Delta x \, \partial_x (f_V' \partial_x \hat{S})(x_k,t) + O(\Delta x^2)$$

$$(3.4\text{-}14)$$

since the second derivatives of Hermite cubic interpolates are $O(\Delta x^2)$ approximations (Prenter, 1975, Section 3.4). Thus, to within $O(\Delta x^2)$, the collocation scheme (3.4-12) corresponds to a parabolic extension including dissipative effects. It is clear in this case that the dissipation vanishes as $\Delta x \to 0$, guaranteeing the shock condition in a numerically consistent fashion.

These results demonstrate the applicability of collocation to multiphase flows in porous media. In particular, upstream collocation as implemented in this section offers an easily coded approach to overcoming the difficulties that Sincovec (1977) reported in solving the purely hyperbolic Buckley-Leverett saturation equation. For this reason the technique should also prove useful in solving the types of near-hyperbolic equations encountered in more complicated models of porous-media flows.

3.5. A gas-flow equation.

The last problem that we shall examine in this chapter is a nonlinear equation governing the flow of ideal gases in porous media. This equation is a simple parabolic analog of the balance law for total fluid mass in a miscible gas flood. The latter equation arises when we sum the species transport equation (1.5-3) over all species i, getting an equation of the form

$$A \, \partial_t \rho - \partial_x (T_T \, \partial_x p_V + \gamma \, \partial_x D - T_L \, \partial_x p_{CVL}) = 0$$

$$(3.5\text{-}1)$$

where $T_T = T_V + T_L$ is the total fluid transmissibility, $\gamma = (T_V \rho^V + T_L \rho^L)g$, and D is depth below some datum. If we assume that the cross-sectional area A is uniform, that only the vapor phase is present, and that gravitational effects are absent, we find

$$\phi \, \partial_t \rho^V - \partial_x (\Lambda_V \, \rho^V \, \partial_x p_V) = 0$$

$$(3.5\text{-}2)$$

where $\Lambda_V = T_V / \rho^V A$ is the vapor mobility. If the vapor is an ideal gas, then it obeys the equation of state $\rho^V = p_V / RT$, where RT is uniform and constant in an isothermal flow field. Using this law and defining a scaled time $\tau = \Lambda_V t / \phi$ reduces equation (3.5-2) to

$$\partial_\tau p = \partial_x (p \, \partial_x p)$$

$$(3.5\text{-}3)$$

where $p = p_V / p_V^{ref}$ is a dimensionless, scaled pressure.

Aronofsky and Jenkins (1951) treated this problem numerically before electronic digital computers were widely available. They used explicit, spatially centered differences to solve several initial boundary-value problems on punch card machines, comparing their results with those of an electric analog device. Let us discuss approximate solutions to equation (3.5-3) on $\Omega \times \Theta = [0,1] \times [0,\infty)$ using orthogonal and upstream collocation along with the auxiliary data

$$p(0,\tau) = 5, \quad \tau \geq 0$$

(3.5-4a)

$$p(1,\tau) = 1, \quad \tau \geq 0$$

(3.5-4b)

$$p(x,0) = 1, \quad x \in (0,1)$$

(3.5-4c)

To discretize (3.5-3) in time let us use the following iterative Euler scheme:

$$p^{n+1,m+1}(x) - p^n(x) = \Delta\tau[\partial_x p^{n+1,m}(x)\ \partial_x p^{n+1,m+1}(x)$$

$$+ p^{n+1,m}(x)\ \partial_x^2 p^{n+1,m+1}(x)]$$

(3.5-5)

Here $p^n(x) = p(x,n\Delta\tau)$ and $p^{n+1,m}$ stands for the m-th iterate of the unknown pressure p^{n+1}. For the spatial discretization let us use the Hermite representation

$$p^n(x) \simeq \hat{p}^n(x) = \sum_{\ell=1}^{M} [\Pi_\ell^n\ H_{0,\ell}(x) + \Pi'^n_\ell\ H_{1,\ell}(x)]$$

(3.5-6)

Substituting (3.5-6) into (3.5-5) and collocating gives

$$\hat{p}^{n+1,m+1}(x_k) - \hat{p}^n(x_k) = \Delta\tau [\partial_x \hat{p}^{n+1,m}(x_k^*) \; \partial_x \hat{p}^{n+1,m+1}(x_k^*)$$

$$+ \; \hat{p}^{n+1,m}(x_k) \; \partial_x^2 \hat{p}^{n+1,m+1}(x_k)]$$

(3.5-7)

As before, the arguments x_k^* stand for upstream collocation points. These points appear in the term that looks "convective" by analogy with the convection-dispersion equation in Section 3.2. The rationale for upstreaming this term is to augment the second-order space derivative by an error $O(\Delta x)$. Huyakorn and Pinder (to appear, Chapter 9) show that, in finite differences, weighting the first-order term in this way is precisely equivalent to using upstream-weighted transmissibilities, a common practice in the petroleum industry. While there is apparently no need for upstream weighting in the single equation (3.5-3), for multi-phase flows an equation having this form is coupled to one or more hyperbolic or nearly hyperbolic transport equations. In these cases upstream weighting is necessary to ensure that the system converges to the physically correct solution.

Figure 3-7 shows the nodal values of the solution to (3.5-3) with auxiliary data (3.5-4) using orthogonal collocation ($x_k^* = x_k$) with $\Delta x = 0.1$ m, $\Delta\tau = 10^{-4}$ m^{3}s^2/kg. The figure also displays curves showing a solution to the same equation using an implicit version of the finite-difference scheme presented by Aronofsky and Jenkins. The difference scheme uses $\Delta x = 0.025$ m, $\Delta\tau = 10^{-4}$ m^{3}s^2/kg, values for which further mesh refinement yields improvements of less than 0.1 percent in the numerical solution. In this case orthogonal collocation using a slightly coarser partition gives results that agree quite well with the finite-difference solution.

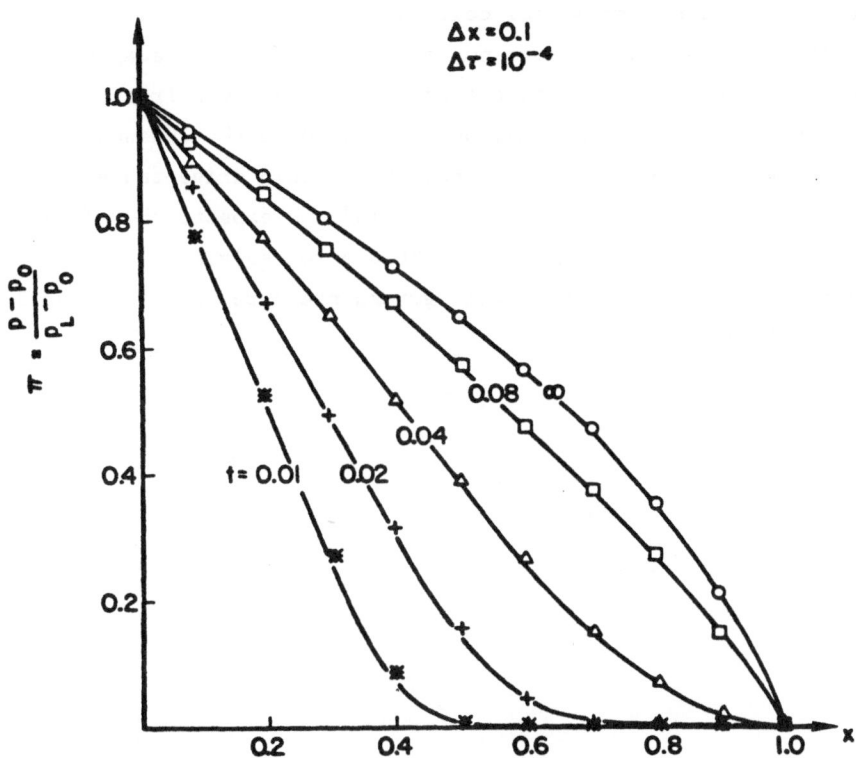

Figure 3-7.   Solutions to the gas-flow problem using
orthogonal collocation.   The smooth curves
represent finite-difference solutions on a fine
grid.

Figure 3-8 is a similar plot for upstream collocation with the upstream points $\xi_k{}^* = \xi_k - 0.4$ in local element coordinates. The solution shows slight smearing in the early profile, with accuracy comparable to that of orthogonal collocation at later times. Upstream collocation in this problem does not effect the kinds of qualitative differences in solution structure observed in the steep-front problems discussed earlier. This fact may be attributed to the absence, except at very early times, of steep gradients that would drive the artificially dissipative error term. It seems fair to expect, therefore, that upstream collocation is a suitable technique for solving nonlinear parabolic equations of the type that govern pressure distributions in multiphase flows.

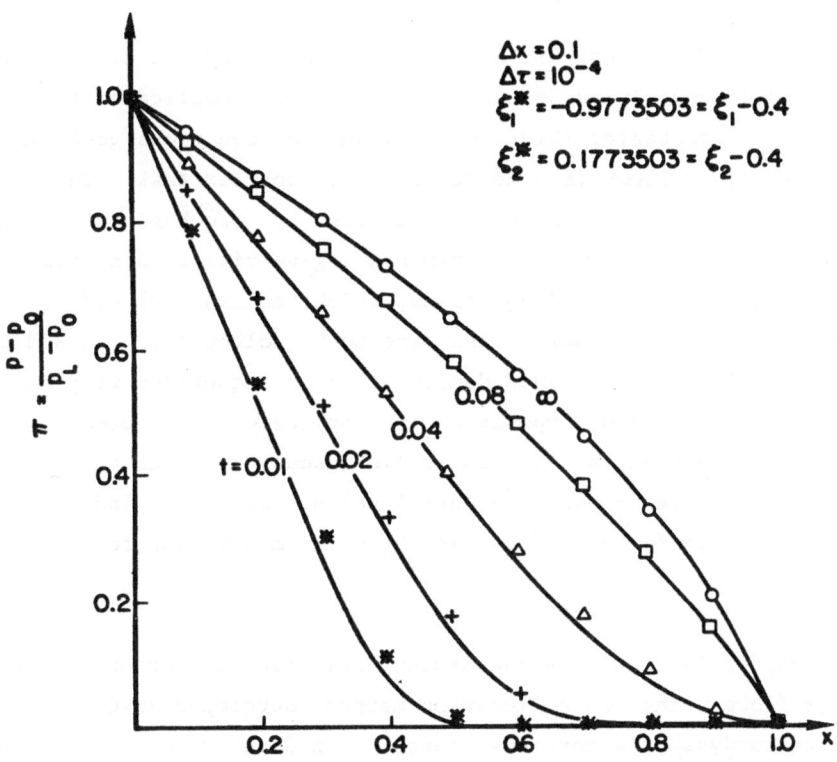

Figure 3-8.   Solutions to the gas-flow problem using
upstream collocation.   The smooth curves
represent finite difference solutions on
a fine grid.

CHAPTER FOUR
MODELING COMPOSITIONAL FLOWS

Porous-media flows that are strongly influenced by the effects of changing fluid compositions are considerably more complicated than the flows examined in Chapter Three, and solving the equations governing such compositional flows is a correspondingly complex task. There are two main reasons for this increase in difficulty. To begin with, the transport laws for a system of N components give rise to a system of N coupled partial differential equations, which require more effort simply because of the larger number of unknowns to be solved for. In addition, the flow coefficients and fluid densities in the equations vary in response to pressure and composition in complicated ways governed by the thermodynamic constraints, as Chapter Two discusses. Thus the nonlinearity of the system imposes another level of hardship. Indeed, the complexity of compositional flows is the prime motivation for modeling them numerically.

This chapter introduces a one-dimensional simulator of compositional reservoir flows using the collocation methods developed in Chapter Three and the thermodynamic algorithms presented in Chapter Two. After developing the numerical techniques used in the simulator, we shall compare the formulation of the collocation-based code with several major compositional simulators reported in the petroleum engineering literature. Section 4.3 examines a set of sample problems using the collocation method. The scheme described below treats one-dimensional flows, and the sample problems involve fluid systems with three or fewer components. While these simplifications would hinder the application of this particular code in actual oilfield operations, the efficacy of the methods in the cases examined here demonstrates the applicability of the overall approach to more ambitious, industrial-scale codes.

4.1. Formulation of the model.

The model uses an implicit pressure - explicit composition scheme to solve the transport equations. This formulation entails solving an overall fluid balance coupled with N - 1 species balance equations. For simplicity let us assume that gravity has no effect on the flow, although this restriction is not essential, and following common practice in compositional modeling let us neglect hydrodynamic dispersion.

Summing the species transport equations (1.5-3) over all N components and using the constraints (1.5-5) gives an overall fluid balance equation:

$$A \, \partial_t \rho = \partial_x (T_T \, \partial_x p_V - T_L \, \partial_x p_{CVL}) = 0$$

$$(4.1-1)$$

where, as in Section 3.5, $T_T = T_V + T_L$ is the total fluid transmissibility. This leaves N - 1 independent species balance equations,

$$A \, \partial_t (\rho \omega_i) = \partial_x (T_i \, \partial_x p_V - T_L \omega_i^L \, \partial_x p_{CVL}) = 0,$$

$$i = 1, \ldots, N-1$$

$$(4.1-2)$$

where $T_i = T_V \omega_i^V + T_L \omega_i^L$. We can regard (4.1-1) as an equation for the pressure p_V, using (4.1-2) to solve for the species mole fractions $\{\omega_i\}_{i=1}^{N-1}$. The restrictions (1.5-5) on mole fractions and saturations, the thermodynamic constraints, and the constitutive laws close the system except for auxiliary data and geometry, as discussed in Section 1.5.

The system of equations comprising (4.1-1) and (4.1-2) is nonlinear, since the coefficients depend rather strongly on the pressure and overall composition of the system. Cast in the implicit pressure - explicit composition form, the equations appear as a parabolic equation in p_V, paralleling the gas-flow equation of Section 3.5, coupled to N - 1 hyperbolic equations in $\omega_1, \ldots, \omega_{N-1}$, each paralleling the Buckley-Leverett equation. The coupling through the thermodynamic and constitutive relationships is pronounced in many practical problems, and under these circumstances it is essential to guarantee consistency among all dependent variables and coefficients at every instant. This observation motivates a solution scheme that iterates over all of the transport equations at each time step in the discrete analog.

Temporal discretization.

To discretize equation (4.1-1) in time let us use an implicit finite-difference scheme having the following Newton-like iterative structure:

$$\rho^{n+1,m} + (\Delta\rho/\Delta p_V)^{n+1,m} \; \delta p_V^{n+1,m+1} - \rho^n$$

$$= \tau \; \partial_x [T_T^{n+1,m} \; \partial_x (p_V^{n+1,m} + \delta p_V^{n+1,m+1})$$

$$- T_L^{n+1,m} \; \partial_x p_{CVL}^{n+1,m}]$$

$$(4.1-3)$$

where $\tau = \Delta t/A$. In this equation, the notation $(\bullet)^n$ stands for the value of the quantity $(\bullet)$ at the known time level n Δt after numerical convergence of the iterations, and $(\bullet)^{n+1,m}$ stands for the most recently computed iterate of $(\bullet)$ at the unknown time level $(n + 1)\Delta t$. The quantity $\delta p_V^{n+1,m+1}$ is the correction to the pressure at the unknown iteration level m + 1, giving $p_V^{n+1,m+1} = p_V^{n+1,m} + \delta p_V^{n+1,m+1}$. The factor $(\Delta\rho/\Delta p_V)^{n+1,m}$ is an approximation to the derivative of fluid density ρ with respect to p_V, determined by a difference quotient

$$(\Delta\rho/\Delta p_V) = [\rho(\omega_1,\ldots,\omega_{N-1}, \ p_V + \Delta p_V)$$

$$- \ \rho(\omega_1,\ldots,\omega_{N-1}, \ p_V)]/\Delta p_V$$

$$(4.1-4)$$

In practice, using $\Delta p_V \simeq 10^2$ Pa gives adequate results.

As Section 4.2 explains, the formulation (4.1-3) parallels that presented by Nghiem et al. (1981) in their finite-difference simulator, the most salient difference being the approximation used for $\Delta\rho/\Delta p_V$. The temporal approximation resembles a Newton-Raphson iterative scheme, except that the right side of (4.1-3) neglects derivatives of the flow coefficients and capillary term with respect to pressure. Although the resulting matrix multiplying the unknowns $\delta p_V^{n+1,m+1}$ is only an approximation of the true Jacobian matrix, experience shows that the scheme converges well while avoiding much of the expense required to compute the true Jacobian.

Given iterates $p_V^{n+1,m+1}$ from the pressure equation, we can update each mole fraction $\omega_1,\ldots,\omega_{N-1}$ using the Euler-like scheme

$$\Delta_t(\rho\omega_i)^{n+1,m+1} = \tau \ \partial_x[T_i^{n+1,m} \ \partial_x p_V^{n+1,m+1}$$

$$- \ (T_L\omega_i^L)^{n+1,m} \ \partial_x p_{CVL}^{n+1,m}]$$

$$i = 1,\ldots,N-1$$

$$(4.1-5)$$

This becomes an equation for the time increment $\Delta_t\omega_i^{n+1,m+1} = \omega_i^{n+1,m+1}$

$-\omega_i^n$ if we expand the left side of (4.1-5) and rearrange:

$$\Delta_t \omega_i^{n+1,m+1} = \{\tau\ \partial_x[T_i^{n+1,m}\ \partial_x p_V^{n+1,m+1}$$

$$- (T_L\omega_i^L)^{n+1,m+1}\ \partial_x p_{CVL}^{n+1,m}] - \omega_i^n\ \Delta_t \rho^{n+1,m+1}\}/\rho^{n+1,m+1}$$

$$(4.1-6)$$

Equation (4.1-6) calls for the values of $\rho^{n+1,m+1}$, which are available from the latest iteration of the pressure equation as

$$\rho^{n+1,m+1} = \tau\ \partial_x(T_T^{n+1,m+1}\ \partial_x p_V^{n+1,m+1} - T_L^{n+1,m}\ \partial_x p_{CVL}^{n+1,m})$$

$$+ \rho^n$$

$$(4.1-7)$$

Spatial discretization.

Let us approximate the spatial variations in the unknowns $\delta p_V^{n+1,m+1}$, $\Delta_t \omega_i^{n+1,m+1}$, $i = 1,\ldots,N-1$, using Hermite cubic trial functions like those defined in Chapter Three. Thus, given a uniform partition Δ_M: $0 = \bar{x}_1 < \ldots < \bar{x}_M = x_{max}$ of the spatial domain $\Omega = [0,x_{max}]$,

$$\delta p_V^{n+1,m+1} \approx \delta \hat{p}^{n+1,m+1}$$

$$= \sum_{\ell=1}^{M} [\pi_\ell^{n+1,m+1}\ H_{0,\ell}(x) + \pi'_\ell^{n+1,m+1}\ H_{1,\ell}(x)]$$

$$(4.1-8)$$

and

$$\Delta_t \omega_i^{n+1,m+1} \simeq \Delta_t \hat{\omega}_i^{n+1,m+1}$$

$$= \sum_{\ell=1}^{M} [w_{i,\ell}^{n+1,m+1} H_{0,\ell}(x) + w'_{i,\ell}^{n+1,m+1} H_{1,\ell}(x)]$$

$$(4.1-9)$$

To these polynomial representations for the increments correspond similar expressions for the total pressure and compositions:

$$P_V^{n+1,m+1} \simeq \hat{P}^{n+1,m+1}$$

$$= \sum_{\ell=1}^{M} [\Pi_\ell^{n+1,m+1} H_{0,\ell}(x) + \Pi'_\ell^{n+1,m+1} H_{1,\ell}(x)]$$

$$(4.1-10)$$

$$\omega_i^{n+1,m+1} \simeq \hat{\omega}_i^{n+1,m+1}$$

$$= \sum_{\ell=1}^{M} [W_{i,\ell}^{n+1,m+1} H_{0,\ell}(x) + W'_{i,\ell}^{n+1,m+1} H_{1,\ell}(x)]$$

$$(4.1-11)$$

according to the updating rules

$$\Pi_\ell^{n+1,m+1} = \Pi_\ell^{n+1,m} + \pi_\ell^{n+1,m+1}$$

$$(4.1-12a)$$

$$\Pi'_\ell^{n+1,m+1} = \Pi'_\ell^{n+1,m} + \pi'_\ell^{n+1,m+1}$$

$$(4.1-12b)$$

$$W_{i,\ell}^{n+1,m+1} = W_{i,\ell}^{n+1,m} + w_{i,\ell}^{n+1,m+1}$$

$$(4.1-12c)$$

$$W'_{i,\ell}^{n+1,m+1} = W'_{i,\ell}^{n+1,m} + w'_{i,\ell}^{n+1,m+1}$$

$$(4.1-12d)$$

For such dependent variables as densities and flow coefficients there are several possible spatial approximations. Since we intend to force the finite-element approximations of the governing equations to hold at the 2M - 2 collocation points x_k, perhaps the most naive approximation for a given dependent variable X is one of the form $X(x_k) \approx X(\hat{\omega}_1(x_k), \ldots, \hat{\omega}_{N-1}(x_k), p(x_k))$. There are two disadvantages to using interpolated values of the principal unknowns in this fashion. First, it requires 2M - 2 evaluations of each dependent variable at each iteration, and for densities, especially, each evaluation can involve expensive thermodynamic calculations. Second, and more important, the interpolated values of the variables $(\omega_1, \ldots, \omega_{N-1}, p_V)$ often behave worse than the nodal values as approximations to the physical solution. In fact, a Hermite representation of a monotonic function may have monotonic nodal values without being monotonic over the element interiors (Jensen and Finlayson, 1980). One finds that interpolatory oscillations can give especially troublesome approximations to the fluid mixture density ρ, to which the solution of the pressure equation is rather sensitive.

One way to avoid these oscillations is to interpolate the nodal values of the dependent variables. This approach requires only M evaluations of a given dependent variable at any iteration. For density and its derivatives and the transmissibilities let us use C^0 interpolation in the Lagrange interpolating basis $\{L_\ell\}_{\ell=1}^M$, where

$$
L_\ell(x) = \begin{cases}
(x - \bar{x}_{\ell-1})/\Delta x, & x \in [\bar{x}_{\ell-1}, \bar{x}_\ell] \\
(\bar{x}_{\ell+1} - x)/\Delta x, & x \in [\bar{x}_\ell, \bar{x}_{\ell+1}] \\
0, & x \notin [\bar{x}_{\ell-1}, \bar{x}_{\ell+1}]
\end{cases}
$$

$$(4.1\text{-}13)$$

Thus, for example, given nodal values

$$\rho(\overline{x}_\ell) = \rho(W_{1,\ell}, \ldots, W_{N-1,\ell}, \Pi_\ell)$$

$$(4.1-14)$$

of the fluid mixture density, its interpolate takes the form

$$\rho(x) \cong \hat{\rho}(x) = \sum_{\ell=1}^{M} \rho(\overline{x}_\ell)\, L_\ell(x)$$

$$(4.1-15)$$

Similar approximations hold for $\Delta\rho/\Delta p_V$, T_T, T_L, T_i, and $T_L \omega_i^L$.

Linear interpolation is not appropriate for the capillary pressure, since equations (4.1-3) and (4.1-6) require its second derivative. For example, the pressure equation (4.1-3) contains the term

$$\tau\, \partial_x(- T_L\, \partial_x P_{CVL}) = - \tau(\partial_x T_L)(\partial_x P_{CVL}) - \tau T_L\, \partial_x^2 P_{CVL}$$

$$(4.1-16)$$

A piecewise linear representation for P_{CVL} would give zero for the second term on the right, an approximation lacking physical justification. Instead, on each element $[\overline{x}_\ell, \overline{x}_{\ell+1}]$ we can represent P_{CVL} by a Lagrange quadratic polynomial

$$P_{CVL}(x) \cong \hat{P}_{CVL}(x) = P_{CVL}(\overline{x}_\ell)\, Q_{\ell,1}(x) + P_{CVL}(\overline{x}_{\ell+\frac{1}{2}})\, Q_{\ell,2}(x)$$

$$+ P_{CVL}(\overline{x}_{\ell+1})\, Q_{\ell,3}(x)$$

$$(4.1-17)$$

Here, $\overline{x}_{\ell+\frac{1}{2}} = (\overline{x}_\ell + \overline{x}_{\ell+1})/2$, and

$$Q_{\ell,1}(x) = (x - \bar{x}_{\ell+\frac{1}{2}})(x - \bar{x}_{\ell+1})/(\tfrac{1}{2}\Delta x^2)$$

(4.1-18a)

$$Q_{\ell,2}(x) = - (x - \bar{x}_\ell)(x - \bar{x}_{\ell+1})/(\tfrac{1}{4}\Delta x^2)$$

(4.1-18b)

$$Q_{\ell,3}(x) = (x - \bar{x}_\ell)(x - \bar{x}_{\ell+\frac{1}{2}})/(\tfrac{1}{2}\Delta x^2)$$

(4.1-18c)

This piecewise quadratic representation can have nonzero first and second derivatives over the interior of each element. To compute $P_{CVL}(\bar{x}_{\ell+\frac{1}{2}})$, we can simply use linear interpolation to estimate its arguments $S_V(\bar{x}_{\ell+\frac{1}{2}})$ and $\sigma(\bar{x}_{\ell+\frac{1}{2}})$ based on their nodal values.

Auxiliary conditions.

The N coupled pressure and composition equations require N initial conditions, specified numerically as follows:

$$\hat{p}(x,0) = p_0(x), \quad x \in (0,x_{max})$$

(4.1-19a)

$$\hat{\omega}_i(x,0) = \omega_{i,0}(x), \quad x \in (0,x_{max}), \ i = 1,\ldots,N-1$$

(4.1-19b)

where the functions on the right sides of these equations lie in the Hermite cubic space $H_3(\Delta_M)$. The restrictive equations and the thermodynamic constraints then determine the initial phase compositions, densities, and saturations (Lake et al., 1981; Pope et al., 1982).

For boundary data, the model admits either specified injection rate or specified pressure at $x = 0$ and specified pressure at the producing end, $x = x_{max}$. Also, the model assumes that the composition of the injected mixture is known. These conditions translate to

$$T_T \ \partial_x \hat{p}(0,t) = - \ q(t) \quad \text{or} \quad \hat{p}(0,t) = p_q(t), \ t \geq 0$$

$$(4.1\text{-}20a)$$

$$\hat{p}(x_{max}) = p_r(t), \quad t \geq 0$$

$$(4.1\text{-}20b)$$

$$\hat{\omega}_i(0,t) = \omega_q(t), \ t \geq 0, \quad i = 1,\ldots,N\text{-}1$$

$$(4.1\text{-}20c)$$

(Price and Donohue, 1967). To ensure that the discrete problem is algebraically closed, let us also impose the artificial conditions

$$\partial_x \hat{\omega}_i(x_{max},t) = 0, \ t \geq 0, \ i = 1,\ldots,N\text{-}1$$

$$(4.1\text{-}20d)$$

at the producing end of the reservoir. These last conditions correspond to the prohibition of dispersive flux at $x = x_{max}$, a stipulation that would be correct were hydrodynamic dispersion included in the species transport equations (Lake et al., 1981). As with the Buckley-Leverett problem of Section 3.4, one could circumvent this artificial condition by adding another collocation point in the rightmost element.

Solving the system.

Substituting the spatial approximations defined above into the time-differenced equations (4.1-3) and (4.1-6) and collocating gives a set of linear algebraic equations at each iteration level. For the pressure equation, we get

$$(\widehat{\Delta\rho/\Delta p_V})^{n+1,m}(x_k) \ \delta\hat{p}^{n+1,m+1}(x_k)$$

$$- \tau\{\partial_x \hat{T}_T^{n+1,m}(x_k^*) \; \partial_x [\delta\hat{p}(x_k^*)]^{n+1,m+1}$$

$$+ \hat{T}_T^{n+1,m}(x_k) \; \partial_x^2 [\delta\hat{p}(x_k)]^{n+1,m+1}\}$$

$$= - \{\hat{\rho}^{n+1,m}(x_k) - \hat{\rho}^n(x_k) - \tau[\partial_x \hat{T}_T^{n+1,m}(x_k^*) \; \partial_x \hat{p}^{n+1,m}(x_k^*)$$

$$+ \hat{T}_T^{n+1,m}(x_k) \; \partial_x^2 \hat{p}^{n+1,m}(x_k) - \partial_x \hat{T}_L^{n+1,m}(x_k) \; \partial_x \hat{P}_{CVL}^{n+1,m}(x_k)$$

$$- \hat{T}_L^{n+1,m}(x_k) \; \partial_x^2 \hat{P}_{CVL}^{n+1,m}(x_k)]\}, \qquad k = 1,\ldots,2M-2$$

$$(4.1\text{-}21)$$

where, as in Chapter Three, x_k^* represents an upstream collocation point, assigned to the appropriate terms by analogy with the gas flow equation examined in Section 3.5. Similarly, the composition equations yield algebraic analogs of the form

$$\Delta_t \hat{\omega}_i^{n+1,m+1} = \{\tau[\partial_x \hat{T}_i^{n+1,m}(x_k^*) \; \partial_x \hat{p}^{n+1,m+1}(x_k^*)$$

$$+ \hat{T}_i^{n+1,m}(x_k) \; \partial_x^2 \hat{p}^{n+1,m+1}(x_k)$$

$$- \partial_x (\widehat{T_L \omega_i^L})^{n+1,m}(x_k) \; \partial_x \hat{P}_{CVL}^{n+1,m}(x_k)$$

$$- (\widehat{T_L \omega_i^L})^{n+1,m}(x_k) \; \partial_x^2 \hat{P}_{CVL}^{n+1,m}(x_k)]$$

$$- \hat{\omega}_i^n(x_k)\Delta_t \hat{\rho}^{n+1,m+1}(x_k)\} / \hat{\rho}^{n+1,m+1}(x_k), \qquad i = 1,\ldots,N-1.$$

$$(4.1\text{-}22)$$

where the assignment of upstream terms follows analogies with the convection-dispersion equation equation of Section 3.3 and the Buckley-Leverett equation of Section 3.4.

The flowchart drawn in Figure 4-1 illustrates the model's overall logic. This scheme calls for the solution of the pressure equation (4.1-21) at each iteration level, a task requiring the inversion of a linear system

$$M^{n+1,m} \, \underset{\sim}{\pi}^{n+1,m+1} = - \, \bar{\underset{\sim}{R}}^{n+1,m}$$

$$(4.1-23)$$

where $\underset{\sim}{\pi}^{n+1,m+1}$ signifies the column vector containing the increments $\{\pi_\ell^{n+1,m+1}, \pi'^{n+1,m+1}_\ell\}_{\ell=1}^M$, $M^{n+1,m}$ is the Jacobian-like matrix containing the coefficients of the left side of (4.1-21), and $\bar{\underset{\sim}{R}}^{n+1,m}$ is the vector of residuals at the most recently computed iteration level, given by the right side of (4.1-21). The code solves equation (4.1-23) directly using a direct solution algorithm for asymmetric banded matrices.

After updating the pressure iterates, the model solves the composition equations (4.1-22). These equations have the form

$$B \, \underset{\sim}{w}_i^{n+1,m+1} = \underset{\sim}{r}_i^{n+1,m}, \qquad i = 1,\ldots,N-1$$

$$(4.1-24)$$

where $\underset{\sim}{w}_i^{n+1,m+1}$ is the column vector of unknown increments $\{w_{i,\ell}^{n+1,m+1},$ $w'^{n+1,m+1}_{i,\ell}\}_{\ell=1}^M$, B is the matrix of Hermite cubic interpolation coefficients giving the values of $\Delta_t \hat{\omega}_i(x_k)$, $k = 1,\ldots,2M-2$, in terms of nodal values, and $\underset{\sim}{r}_i^{n+1,m}$ is the column vector of right sides from equation (4.1-22). Since B is constant we need only invert it once, during initialization, reducing (4.1-24) to the matrix multiplications $\underset{\sim}{w}_i^{n+1,m+1} = B^{-1} \underset{\sim}{r}_i^{n+1,m}$ at each iteration.

After computing the iterates $\hat{p}^{n+1,m+1}$ and $\hat{\omega}_i^{n+1,m+1}$, $i = 1,\ldots,N-1$, the model performs the thermodynamic calculations, using the methods of

138

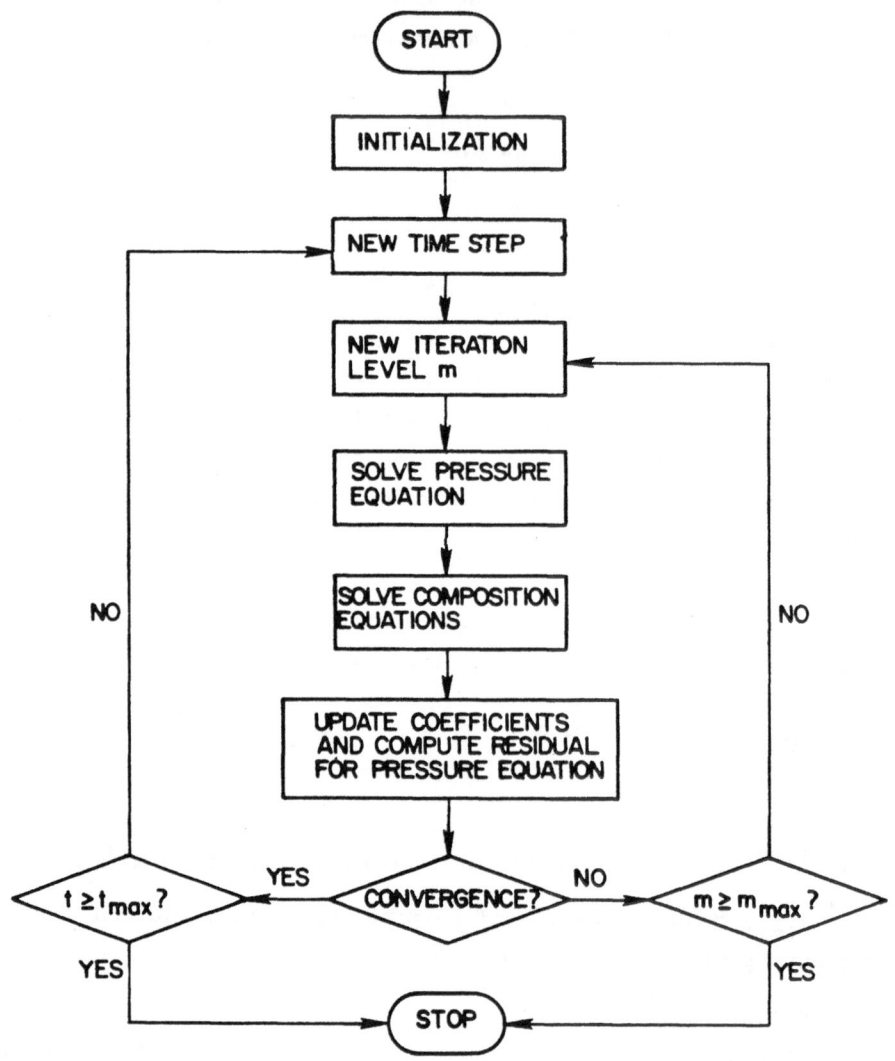

Figure 4-1. Flowchart of the overall structure of the
compositional simulator.

Section 2.3 to calculate vapor-liquid equilibria, and evaluates the constitutive variables. To test for numerical convergence at each iteration, the code checks criteria of the form

$$\max_{\ell} \{|\hat{p}^{n+1,m+1}(\bar{x}_\ell) - \hat{p}^{n+1,m}(\bar{x}_\ell)|\} < \varepsilon_p$$

$$\text{(4.1-25a)}$$

$$\max_{\ell} \{|\hat{\omega}_i^{n+1,m+1}(\bar{x}_\ell) - \hat{\omega}_i^{n+1,m}(\bar{x}_\ell)|\} < \varepsilon_\omega$$

$$\text{(4.1-25b)}$$

In practice, ε_p = 5.0 kPa and ε_ω = 0.001, values comparable to those imposed by Coats (1980).

Stability and iterative convergence.

There appears to be no simple a priori way to compute the stability limits of this coupled system. Taken alone, the pressure equation should admit large time steps owing to its implicitness. The composition equations, being explicit, should be vulnerable to instability when the time steps allow the injection of more fluid than can fit in an element at the prevailing local pressure. This criterion parallels the Courant-number condition for explicit finite-difference schemes. In practice, however, large changes in overall fluid compressibility accompanying changes in composition and pressure can make the system unstable at smaller time steps. This sensitivity typically arises when the vapor and liquid differ greatly in density and exchange little mass.

The code allows for the possibility that the size of stable time steps will increase with time, a phenomenon that manifests itself through convergence of the iterative scheme in progressively fewer iterations. When the system converges in one iteration, the model automatically doubles its time step, leading to more efficient use of the computational effort. Occasionally the opposite happens. As the

pressure decreases below a bubble point, for example, the scheme demands
a reduction in time step. The code therefore allows up to five halvings
of the time step before aborting the calculations. In most examples the
code regains stability after at most two such halvings in any single
time step.

As the coupled system converges during each time step the residual
from the pressure equation decreases at a roughly linear rate. This
rate accords with results plotted by Mansoori (1982) for a finite-
difference simulator of similar design. Figure 4-2 shows a plot of the
logarithms of successive residual norms for the collocation code, corro-
borating the approximately linear asymptotic rate. In theory, an exact
Newton-Raphson scheme for the pressure equation solved alone should
exhibit a quadratic asymptotic rate. Apparently the convergence for the
approximate Newton-Raphson scheme in equation (4.1-21) is slower when
this equation is coupled to other balance equations solved explicitly.

Material balance.

The fact that the transport equations for compositional flows are
essentially statements of mass conservation is a compelling reason to
provide independent checks on material balance. In a Newton-like scheme
such as that used for the pressure equation (4.1-21), norms of the
residual vector $\bar{R}^n$ provide useful checks on the pointwise conservation
of total fluid mass. Since the elements of $\bar{R}^n$ are residuals at the
collcation points,

$$A \int_\Omega \bar{R}^n(x) \ dx \simeq (2A/\Delta x) \sum_{k=1}^{K_c} \bar{R}_k^n,$$

$$(4.1-26)$$

where $K_c = 2M - 2$, renders a Gauss-quadrature approximation to the
global fluid mass conservation during any time step. In terms of this

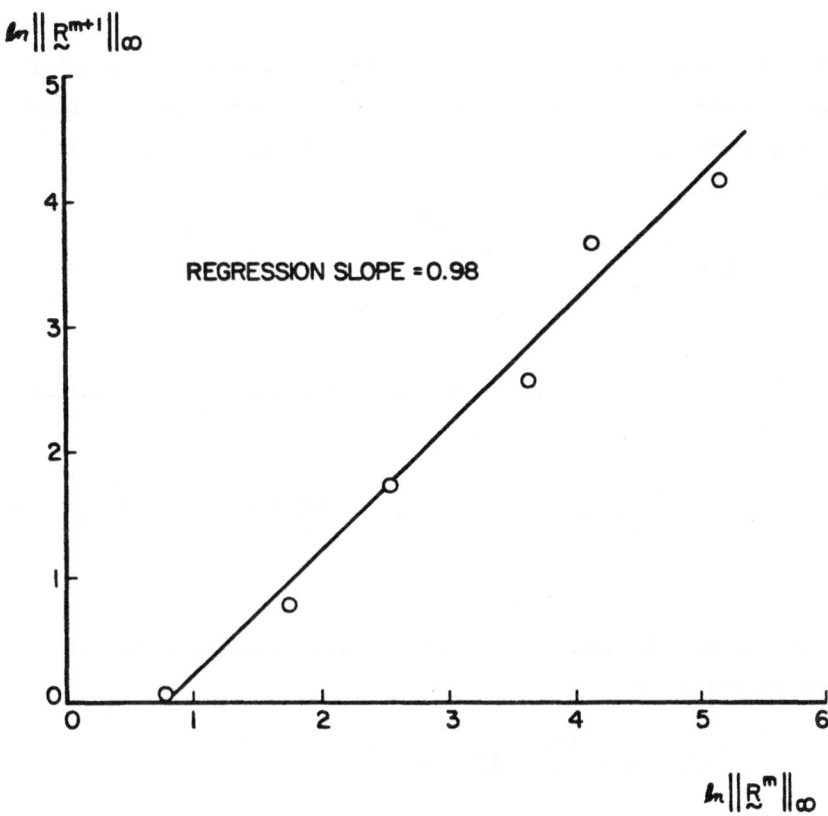

Figure 4-2. Logarithms of successive residual norms
from iterates of the pressure equation for a
three-component flow.

approximate measure, the model typically conserves mass to within a fraction of one percent of the mass present in the system.

Similar approximate measures of global conservation exist for each fluid component i. Rewriting the species balance (4.1-5) using the finite-element representations of its terms gives an equation of the form

$$\Delta_t (\hat{\rho}\hat{\omega}_i)^{n+1} = \tau \, \partial_x \hat{F}_i^{n+1}$$

(4.1-27)

where $\hat{F}_i^{n+1}$ is an abbreviation for the flux terms. Integrating this expression over the spatial domain Ω yields

$$\int_\Omega \hat{\rho}^{n+1}\hat{\omega}_i^{n+1} \, dx - \int_\Omega \hat{\rho}^n\hat{\omega}_i^n \, dx - \tau[\hat{F}_i^{n+1}(x_{max}) - \hat{F}_i^{n+1}(0)] = 0$$

(4.1-28)

We can compute the mass integrals in this equation using three-point Gauss quadratures:

$$\int_\Omega \hat{\rho}\hat{\omega}_i \, dx = \sum_{\ell=1}^{M-1} \int_\Omega \hat{\rho}\hat{\omega}_i \, dx$$

$$\simeq \sum_{\ell=1}^{M-1} \sum_{j=1}^{3} m_j \, \hat{\rho}(x_{j,\ell})\hat{\omega}_i(x_{j,\ell})$$

(4.1-29)

where the numbers m_j are weights and the values $x_{j,\ell}$ are the three-point Gauss points in the element $[\bar{x}_\ell, \bar{x}_{\ell+1}]$. The model typically conserves the masses of individual species to within errors, measured in this way, of a fraction of a percent of the mass of the species present.

4.2. Connections to other compositional models.

The compositional simulator described in the previous section is related in structure to other implicit pressure - explicit composition models. Nolen (1973) and Kazemi et al. (1978), for example, used this general approach in models in which tabulated correlations provide the basis for vapor-liquid equilibrium calculations. Nghiem et al. (1981) extend this formulation to include equation-of-state techniques for imposing thermodynamic constraints. In approximating the Jacobian to the pressure equaticn, Nghiem et al. neglect the responses of saturations and phase compositions to pressure changes, setting

$$\Delta p / \Delta p_V \simeq \phi \sum_\alpha S_\alpha \, \partial_{p_V} \rho^\alpha$$

$$(4.2\text{-}1)$$

where α ranges over all fluid phases. Their approximate Jacobian, unlike the matrix $\mathbf{M}$ in equation (4.1-23), is symmetric and diagonally dominant, allowing some computational savings over asymmetric approximations. There seems to be no physically motivated contrivance that will bestow these properties on the approximate Jacobian $\mathbf{M}$ arising in collocation. Mansoori (1982) proposes an improvement to (4.2-1) including density-driven saturation changes but still ignoring the responses of phase compositions. This modification enhances the diagonal dominance of the approximate Jacobian for many hydrocarbon fluid systems, but its implications in systems exhibiting retrograde condensation are unclear. The finite-difference approximation (4.1-4) includes the full effects of pressure changes on phase compositions as well as saturations and phase densities, but it is only fair to note that it also requires an additional suite of thermodynamic calculations at each iteration.

The collocation model breaks with tradition in that, in contrast to
other implicit pressure - explicit composition simulators, the code uses
implicit transmissibilities and capillary terms. However, the transmissibilities and other flux-related dependent variables are lagged by one
iteration, so their derivatives with respect to pressure do not
contribute to the approximate Jacobian. Using implicit transmissibilities improves the internal consistency of the numerical soutions at each
time step.

There is another important approach to compositional simulation,
namely, the fully implicit formulation. Models based on this approach
solve the flow equations and the thermodynamic constraints simultaneously and implicitly, a strategy yielding codes that are stable for
large time steps at the expense of considerable computation. Fussell
and Fussell (1979), for example, present a "minimum-variable Newton-
Raphson" (MVNR) technique for choosing the principal iteration variables
in such a code. Their model, based on the Redlich-Kwong equation of
state, assumes explicit flow coefficients and the absence of capillary
effects. The MVNR approach requires the simulator to determine at the
beginning of each time step how many fluid phases exist in each zone of
the spatial grid. This dependence on what is essentially an explicit
saturation pressure imposes some limitations on time steps. Coats
(1980) develops another fully implicit equation-of-state simulator
including implicit flow coefficients and gravitational and capillary
effects. This simulator uses a fixed set of principal iteration
variables and solves the standard saturation-pressure equations at each
iteration, thus avoiding the time-step limitations of the MVNR construction. As Coats discusses, the fully implicit formulation exacts a
penalty in return for its stability: as the number of components grows,
the size of the matrix to be inverted at each iteration grows as well,
demanding rapid increases in computational effort.

Other solution schemes have appeared. Price and Donohue (1967), Roebuck et al. (1969), and Van Quy et al. (1973) all use formulations in which all N mixture compositions $\{\omega_i\}_{i=1}^N$ are principal unknowns. Price and Donohue and Van Quy et al. impose the restriction $\sum_{i=1}^N \omega_i = 1$ to compute the pressure gradient at each node, integrating the gradient from a boundary point having prescribed pressure to calculate interior pressures. Roebuck et al. solve iteratively for pressures consistent with the other thermodynamic properties of the phases. Schemes such as these appear to have found few adherents in the recent literature.

The collocation simulator follows a recent trend in incorporating the effects of phase compositions on relative permeabilities and capillary pressure through the interfacial tension of the fluids. Coats (1980) and Nghiem et al. (1981) propose analytic formulas to model these effects, and Bardon and Longeron (1980) and Amaefule and Handy (1981) have published similar empirical correlations. For relative permeabilities, let us use the equation

$$k_{r\alpha} = \begin{cases} k_{r\alpha}^{max}[R_\sigma \overline{S}_\alpha^2 + (1 - R_\sigma)\overline{S}_\alpha], & S_\alpha \geq S_{\alpha R} \\ \\ 0, & S_\alpha < S_{\alpha R} \end{cases}$$

$$(4.2\text{-}2)$$

for α = V or L, where

$$R_\sigma = (\sigma/\sigma^{ref})^{\frac{1}{4}}$$

$$(4.2\text{-}3)$$

measures the effect of changes in interfacial tension from a reference state and $S_{\alpha R}$ stands for the residual saturation of phase α, α = V or L. Equation (4.2-2) uses a normalized saturation,

$$\bar{S}_\alpha = (S_\alpha - S_{\alpha R})/(1 - S_{VR} - S_{LR})$$

$$(4.2-4)$$

with the residual saturation scaled from its value at the reference state:

$$S_{\alpha R} = R_\sigma \, S_{\alpha R}^{ref}$$

$$(4.2-5)$$

For capillary pressure, let us assume a drainage curve of the form

$$P_{CVL} = (\sigma/\sigma^{ref})(c_0 + c_1 S_L + c_2 S_L^2),$$

$$(4.2-6)$$

choosing the coefficients c_j of the polynomial to force it to pass through a specified value at $S_L = S_{LR}$ and to have a specified value with zero slope at either $S_L = 1 - S_{VR}$ or $S_L = 1$. Equations (4.2-2) through (4.2-6) exhibit the generic properties discussed in Section 1.4. In actual applications these curves would be based on experimental data.

4.3. Examples of compositional flows.

We shall examine three representative problems using the compositional model described above. These problems demonstrate the performance of the simulator and numerically corroborate several design principles based on qualitative features of gas floods. The examples include
an immiscible solution-gas drive, a condensing gas drive, and a vaporizing gas drive.

Example 1: solution-gas drive.

In this example, the reservoir fluids consist of two species, "gas"
and "oil", whose properties we shall compute by using the equation-of-
state parameters for CO_2 and n-decane at 344.26 K for the "gas" and
"oil", respectively. Under the typical assumptions for modeling
solution-gas drives, these species obey a set of thermodynamic
constraints allowing limited mass transfer between the vapor and liquid
phases, namely,

$$\omega_1^V = 1$$

(4.3-1a)

$$\omega_2^V = 1 - \omega_1^V = 0$$

(4.3-1b)

$$\omega_1^L = R_s \rho_1^{ref} / \ (R_s \rho_1^{ref} + \rho_2^{ref})$$

(4.3-1c)

$$\omega_2^L = 1 - \omega_1^L$$

(4.3-1d)

where the index 1 stands for "gas" (CO_2) and 2 for "oil" (n-decane). In equation (4.3-1c), ρ_i^{ref} signifies the molar density of species i at some reference state, and R_s is the dimensionless solution gas-oil ratio, defined as the volume of species 1 at reference conditions dissolved per unit volume of species 2 at reference conditions:

$$R_s = \rho_2^{ref}\omega_1^L/(\rho_1^{ref}\omega_2^L)$$

$$(4.3-2)$$

Let us use $(p^{ref},T^{ref}) = (5 \text{ MPa}, 344.26 \text{ K})$ as the reference state, although in oilfield practice it is more common to choose a pressure and temperature typical of the surface separation facilities.

For isothermal reservoirs it is often reasonable to assume that R_s varies only with pressure. In most cases R_s increases monotonically with pressure up to p^{bub}, the bubble pressure of the oil, above which R_s remains constant (Collins, 1963, Chapter 10). Figure 4-3 shows the simple curve of solution gas-oil ratios used in this example.

Table 4-1 lists parameters characterizing the sample problem. The problem includes capillarity with a drainage capillary pressure curve, but no account is made of compositional effects on the capillary pressure or the relative permeabilities.

Figure 4-4 displays model-predicted profiles of pressure, "gas" composition, and vapor saturation for various times. As expected, the saturation curves in Figure 4-4c indicate that this flood is completely immiscible. There is no displacement of oil beyond the original residual saturation S_{LR} because composition has no effect on interfacial tension. What is worse, the solution-gas drive is quite inefficient in moving the originally mobile oil toward the producing end of the reser-

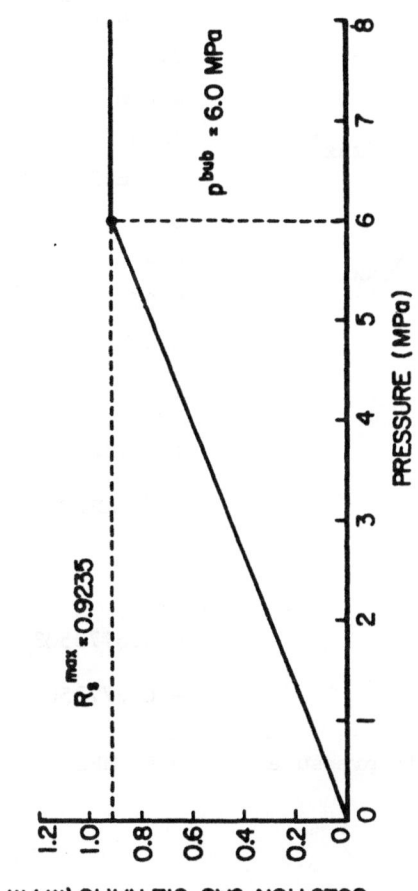

Figure 4-3. Simple curve of solution gas-oil ratios versus pressure.

TABLE 4-1

PARAMETERS USED IN MODELING SOLUTION-GAS DRIVE

Permeability	$1.0 \times 10^{-12} \text{ m}^2$ ($\simeq$ 1 darcy)
Porosity	0.2
Cross-section	1.0 m^2
Reservoir length (x_{max})	10 m
Injection rate	2.0 mol/s
$\omega_1(0,t)$	0.7
$\omega_1(x,0)$ $0 < x \leq x_{max}$	0.4
S_{VR}	0.25
S_{LR}	0.20
$k_{r\alpha,}^{max}$ $\alpha = V, L$	1.0
$P_{CVL}(S_L = S_{LR})$	50 kPa
$P_{CVL}(S_L = 1 - S_{VR})$	5.0 kPa
Δx	1.0 m
Initial Δt	30 s
ξ_1^*	$-$ 0.8773503
ξ_2^*	$-$ 0.2773503
Initial reservoir pressure	5.0 MPa

151

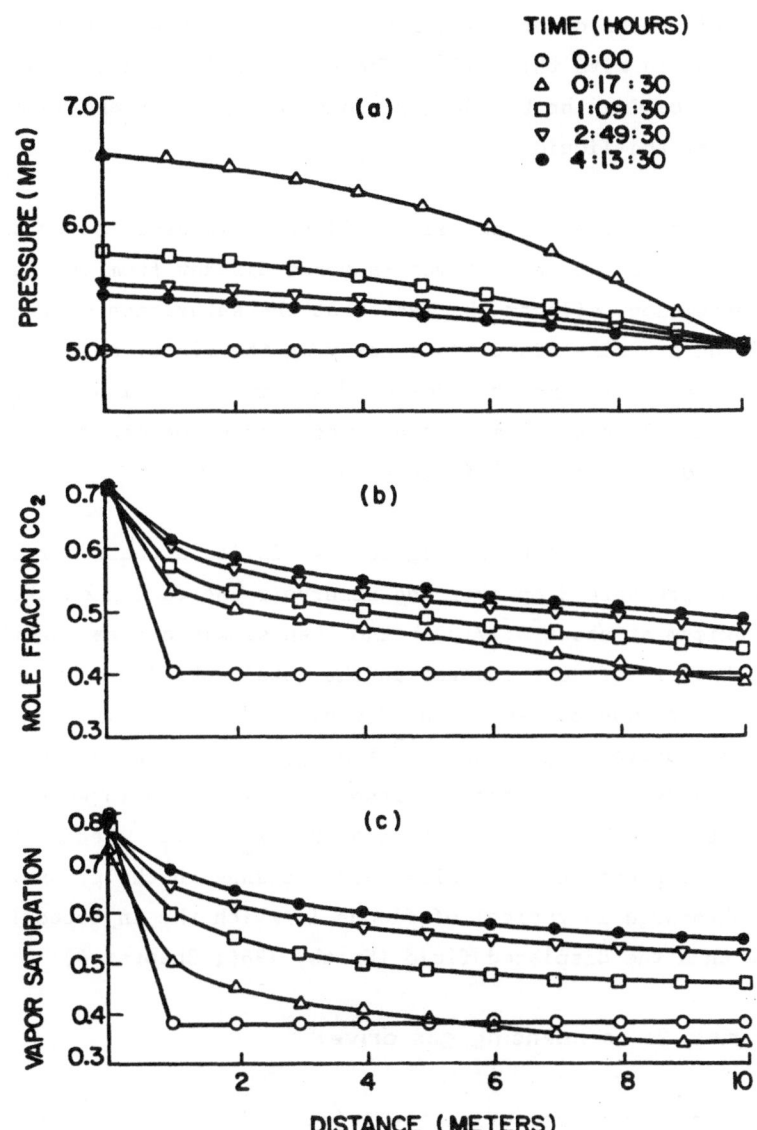

Figure 4-4. Numerical solution to the solution-gas
drive problem.

voir. This inefficiency is a result of the greater mobility of free
"gas" compared with the "oil". The vapor phase bypasses the less mobile
"oil", inhibiting the buildup of a displacing front and leading to slow
recovery of the liquid.

The vapor saturation profile in Figure 4-4c drops below the initial
vapor saturation at early times in zones distant from the injection end.
This phenomenon reflects the fact that the saturation disturbance lags
behind the pressure disturbance early in the flood, so that in the far
field the pressure increase forces "gas" into solution before the wave
of injected fluid can compensate. The phenomenon can occur in more
complicated compositional flows as well.

The pressure profiles in Figure 4-4a follow a characteristic pattern,
the total pressure drop over the reservoir increasing quickly at first
to a maximum at about 0:17:30 h and then slowly decreasing thereafter.
This behavior reflects changes in the total mobility $T_T = T_V + T_L$ of the
fluids during the course of the flood, as drawn in Figure 4-5. At early
times the controlling value $T_T^{min} = \min_x \{T_T(x)\}$ drops as the advancing
vapor lags behind the pressure response. When the injected vapor
finally breaks through the outlet at $x = x_{max}$, T_T^{min} begins to rise and
the overall pressure drop falls. This change in the pressure response
at breakthrough is typical of floods in which the injected fluid is more
mobile than the displaced fluid (Smith, 1966, Chapter 4).

Example 2: condensing gas drive.

In a condensing gas drive the injected fluid is relatively rich in
species having intermediate molecular weights. As the flood progresses,
components of the injected vapor condense into the liquid phase through
continual vapor-liquid contacts, leading to the establishment of a zone

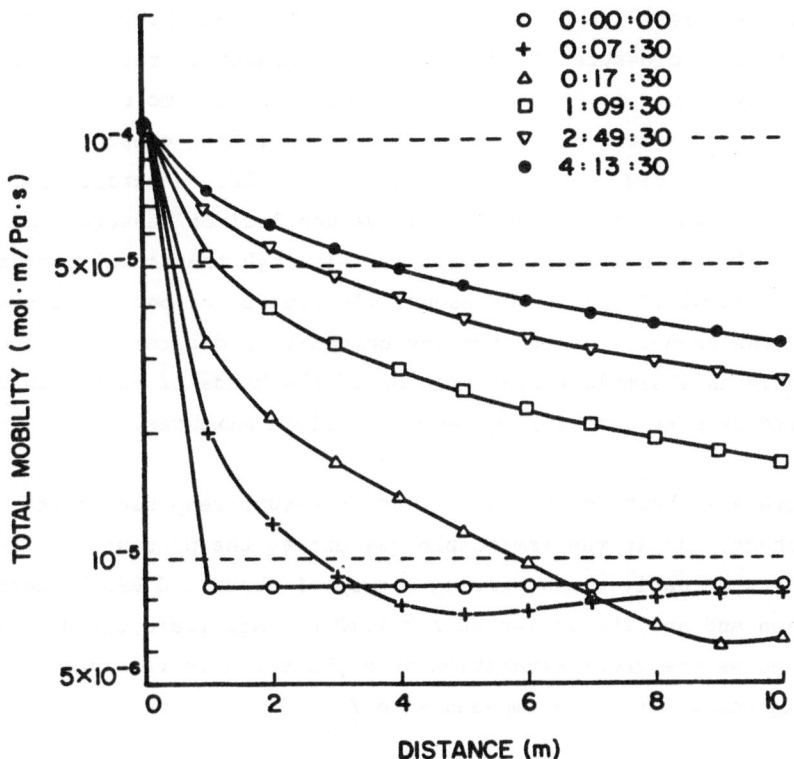

Figure 4-5. Total fluid mobility versus distance
for various times in the solution-gas
drive.

in which the composition of the liquid approaches the critical
composition for the local value of the pressure. The displacing fluid
and the displaced fluid exhibit low interfacial tensions in this zone,
and as a result the displacement is very efficient. The present example
serves as a model of this mechanism.

Table 4-2 lists the parameters defining this example. The reservoir
has an initial pressure of 10 MPa, and the injection rate is 2.0 mol/s.
The injected fluid is 85 mole percent CO_2 with 12.5 mole percent
n-butane and 2.5 mole percent n-decane as enriching components, while
the resident liquid is a saturated mixture of CO_2, n-butane, and
n-decane at 344.26 K. Figure 4-6 shows the *loci* of injected and
resident fluids in composition space. Although the actual composition
of the resident liquid is not especially typical of petroleum reservoir
fluids, the geometry of the ternary composition diagram in this case is
reasonable as a simple representation of the kinds of phase behavior
exhibited by more complicated reservoir fluid mixtures.

Figure 4-7 shows the history of the pressure response to this injection scheme. As in the immiscible gas drive, the pressure in the reservoir rises at early times, giving a drop of about 300 kPa between the
injection end and the outlet at t = 1:30 h. Afterward the pressure
decreases as the vapor saturation throughout the system increases,
lowering the reservoir's impedance to flow.

Figure 4-8 displays the progress of the vapor saturation S_V as the
flood proceeds. Except in the early solutions, the saturation fronts
rise to values significantly larger than the initial maximum vapor
saturation $1 - S_{LR}^0 = 0.80$. Downstream of the saturation peak is a
transition zone over which the fluid system varies continuously in
composition, changing in space from a predominantly vapor system to a

TABLE 4-2

PARAMETERS USED IN MODELING CONDENSING GAS DRIVE

Permeability	$1.0 \times 10^{-2} \ m^2$ ($\simeq$ 1 darcy)
Porosity	0.2
Cross-section	$1.0 \ m^2$
Reservoir length (x_{max})	10 m
Injection rate	2.0 mol/s
$\omega_1(0,t)$	0.85
$\omega_2(0,t)$	0.125
$\omega_1(x,0)$, $0 < x \leq x_{max}$	0.73
$\omega_2(x,0)$, $0 < x \leq x_{max}$	0.05
S_{VR}^o	0.25
S_{LR}^o	0.20
$k_{r\alpha}^{max}$, $\alpha = V,L$	1.0
P_{CVL}	0.0 kPa
Δx	1.0 m
Initital Δt	4:00 min
ξ_1^*	-0.8773503
ξ_2^*	0.2773503
Initial reservoir pressure	10 MPa

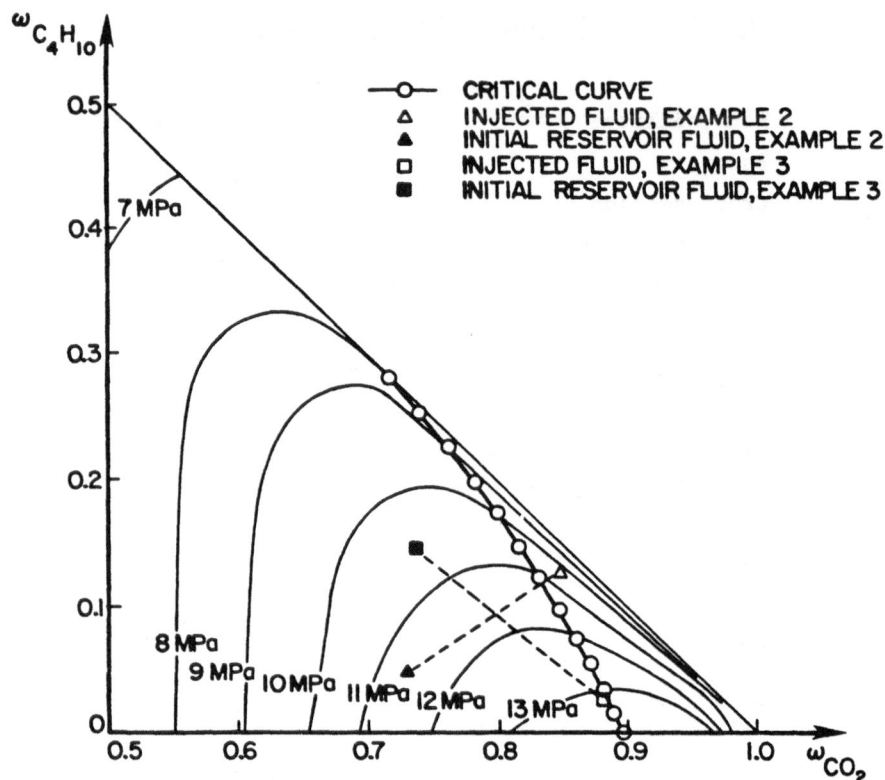

Figure 4-6.   Loci of injected fluids and initial reservoir
              fluids for examples 2 and 3.   The isobars are level
              sets of the saturation pressure dome.

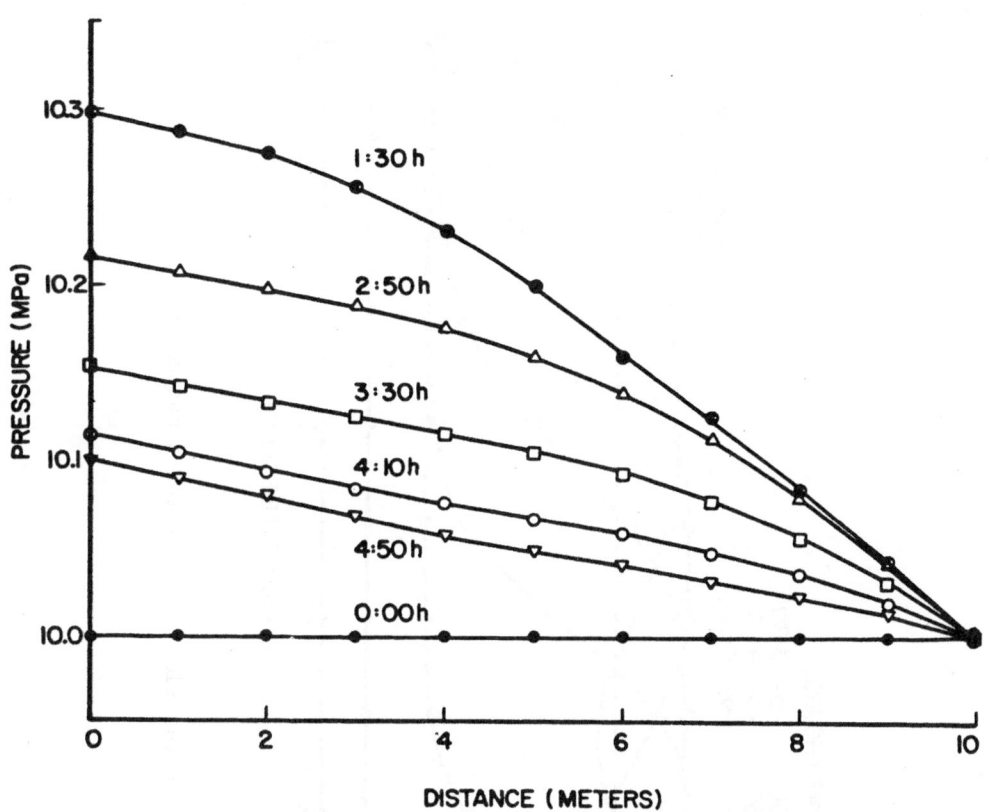

Figure 4-7, History of the pressure response for the condensing-gas drive.

158

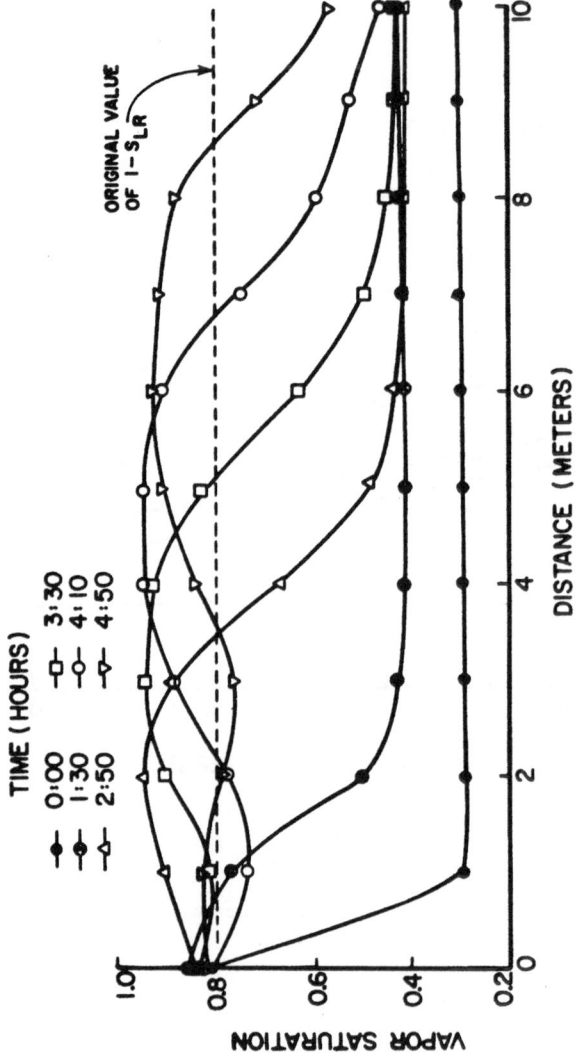

Figure 4-8. Saturation profiles for a condensing gas flood.

predominantly liquid system. This behavior indicates an increase in displacement efficiency over completely immiscible floods, in which no significant lowering of residual liquid saturation occurs. Behind the transition zone the reservoir contains fluid whose composition is close to that of the injected fluid, and therefore the fluid saturations in the tail reflect in large measure the vapor-liquid equilibria of this fluid at the locally prevailing pressures. In particular, as the transition zone passes any point in the reservoir the vapor saturation decreases as the rich injection fluid, relatively unaltered by contact with the original reservoir fluids, occupies the rock's effective pore volume.

The relatively efficient displacement in the transition zone is a consequence of the low interfacial tension in that zone. As Figure 4-9 shows, a wave of depressed interfacial tension traverses the reservoir, the lowest point in the leading edge of the depression corresponding approximately to a point slightly downstream of the peak in vapor saturation. One can account for the lowering of interfacial tension by plotting the early composition history of a spatial point. Figure 4-10 is such a plot for the location x = 2 m. From this figure it is apparent that the liquid composition grows progressively richer in the injected fluid, so that the vapor and liquid in contact become more similar in composition. As a result, the zone of low interfacial tension develops and propagates through the reservoir.

Continuous injection of enriched gas leaves a relatively rich fluid mixture in the swept zones of the reservoir. The dip in S_V behind the transition zone is a symptom of this phenomenon. In actual field projects it is common to displace the enriched injection fluid with a relatively cheap chase gas, often nitrogen or methane, thus lowering both the total cost of the injected fluids and the amount of liquid left

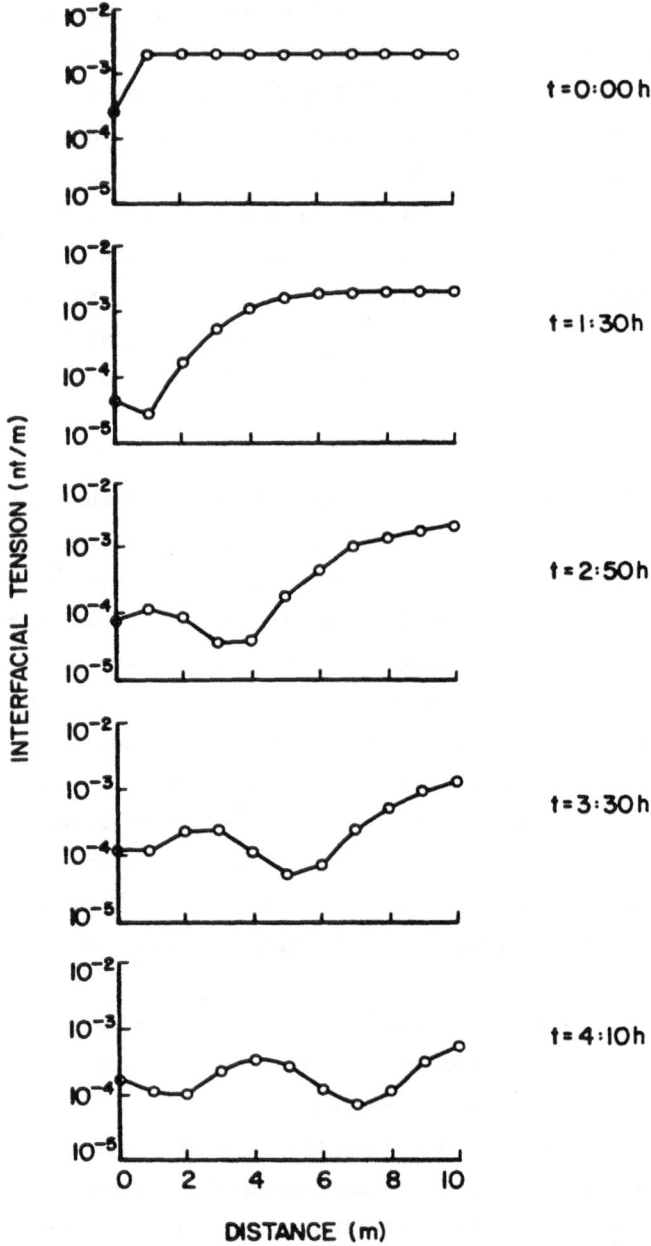

Figure 4-9. Interfacial-tension profile at various times
for the condensing gas drive.

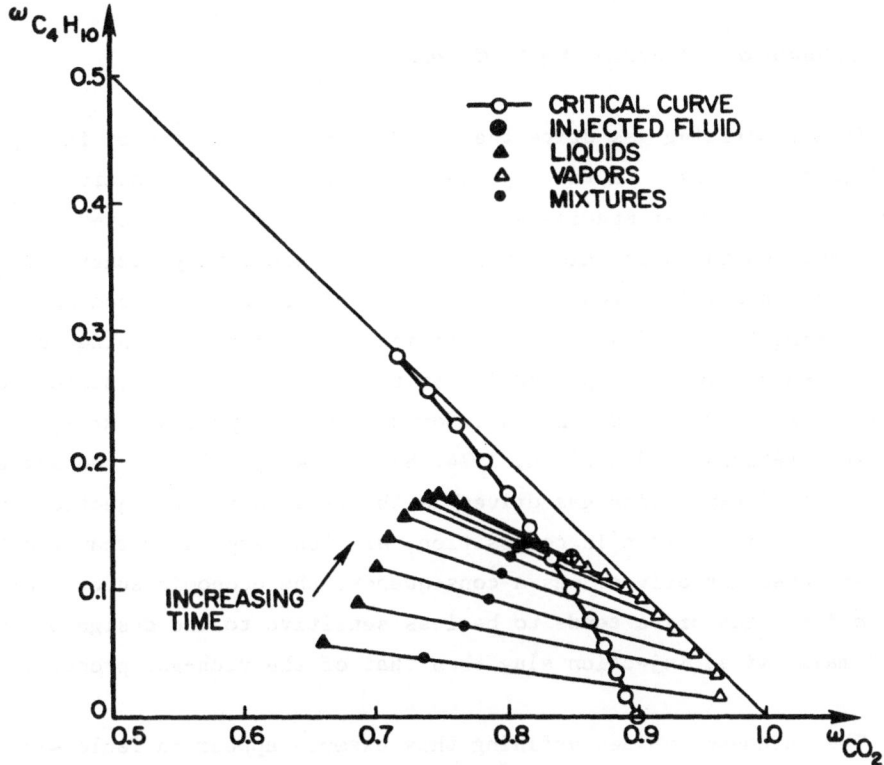

Figure 4-10. Behavior of the fluid mixture composition with
time at x=2 m in the condensing gas flood.

in the reservoir (Mungan and Johansen, 1974). Therefore successful
condensing gas drives in practice leave fairly low non-aqueous liquid
saturations in their wake. The flood simulated in the present example
does not include the injection of a chase gas.

Example 3: vaporizing gas drive.

In a vaporizing gas drive the injected fluid consists of fairly light
molecular species. As the injected vapor contacts the reservoir liquid
some of the heavier species evaporate, enriching the injected fluid.
Continual contacts in this way lead to a progressively richer bank of
driving fluid which eventually acts as a miscible transition zone,
displacing the resident liquid efficiently and leaving behind very low
liquid saturations. Vaporization is the dominant thermodynamic mechanism in many CO_2 floods and is important in high-pressure dry-gas floods
as well (Sandrea and Nielsen, 1974, Section 4.3). One of the attractive
features of vaporizing gas drives is the fact that the injected fluids,
being relatively lean in composition, are less expensive than those used
in enriched gas drives. As a consequence, the economic success of a
vaporizing gas drive tends to be less sensitive to the design of an
optimally sized injection slug than that of the rich-gas process.

The parameter values defining this example appear in Table 4-3. The
injected fluid is 88 mole percent CO_2, two mole percent n-butane, and 10
mole percent n-decane, and, as in Example 2, the reservoir initially
contains a saturated mixture of CO_2, n-butane, and n-decane at 344.26 K.
The loci of these fluids in composition space appear in Figure 4-6. The
initial reservoir pressure is 10 MPa, and the injection rate is 2.0
mol/s.

TABLE 4-3

PARAMETERS USED IN MODELING VAPORIZING GAS DRIVE

Permeability	1.0×10^{-12} m^2 ($\simeq$ 1 darcy)
Porosity	0.2
Cross-section	1.0 m^2
Reservoir length (x_{max})	12 m
Injection rate	2.0 mol/s
$\omega_1(0,t)$	0.88
$\omega_2(0,t)$	0.02
$\omega_1(x,0)$, $0 < x \leq x_{max}$	0.74
$\omega_2(x,0)$, $0 < x \leq x_{max}$	0.14
S_{VR}^{o}	0.25
S_{LR}^{o}	0.20
$k_{r\alpha}^{max}$, $\alpha = V,L$	1.0
P_{CVL} $(S_L = S_{LR})$	5.0 kPa
P_{CVL} $(S_L = 1 - S_{VR})$	5.0 kPa
Δx	1.0 m
Initial Δt	2:00 min
ξ_1^{*}	−0.8773503
ξ_2^{*}	0.2773503
Initial reservoir pressure	10 MPa

Figure 4-11 displays the response of the pressure at various times during the flood. Figure 4-12 shows the history of the vapor saturation front. As with the condensing gas flood, the vapor saturation eventually rises above its initial maximum value $1 - S_{LR}^0 = 0.80$, resulting in very efficient mobilization of the resident liquid. In contrast to the condensing gas flood modeled in Example 2, the vaporizing gas flood does not leave a dip in vapor saturation in its wake. This fact may be attributed to the leanness of the injection fluid. The persistence of low liquid saturations behind the displacing front is attractive for economic and technical reasons, since the vaporizing gas slug is amenable to displacement by a very cheap chase fluid such as water.

As with the condensing gas drive, the displacement front of the vaporizing gas drive is associated with a wave of low interfacial tension. Figure 4-13 plots this wave for various times. Unlike the low-tension wave shown in Figure 4-9, however, the interfacial-tension trough in this example deepens as the flood progresses, suggesting the continued development of a miscible bank. So long as the vaporizing gas slug is sufficiently large, we may expect this process to continue generating its own miscible bank as the flood proceeds.

It is interesting to compare this solution, using upstream collocation, with the numerical solution produced by orthogonal collocation. No such comparison arises in the finite-difference compositional simulators cited above, since all assume upstream weighted transmissibilities. Figure 4-14 shows the saturation profiles corresponding to those in Figure 4-12, the only difference between the two solutions being the choice $x_k^* = x_k$ of collocation points in the model run generating Figure 4-14. The solution using orthogonal collocation fails to capture the effects of developed miscibility, predicting vapor saturations less than $1 - S_{LR}^0$ throughout the flood. It is worth noting that this difference

165

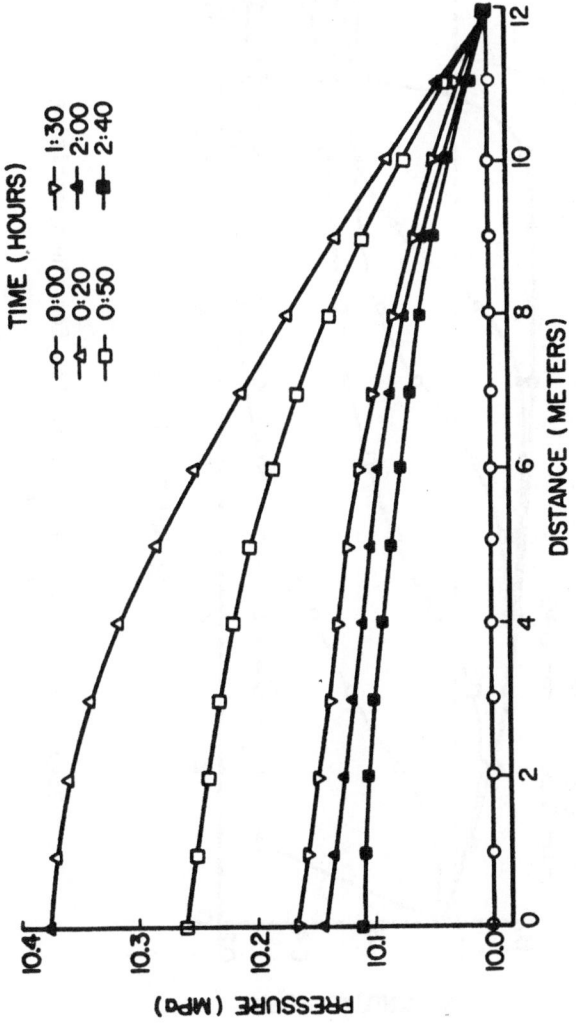

Figure 4-11. History of the pressure response during the vaporizing gas drive.

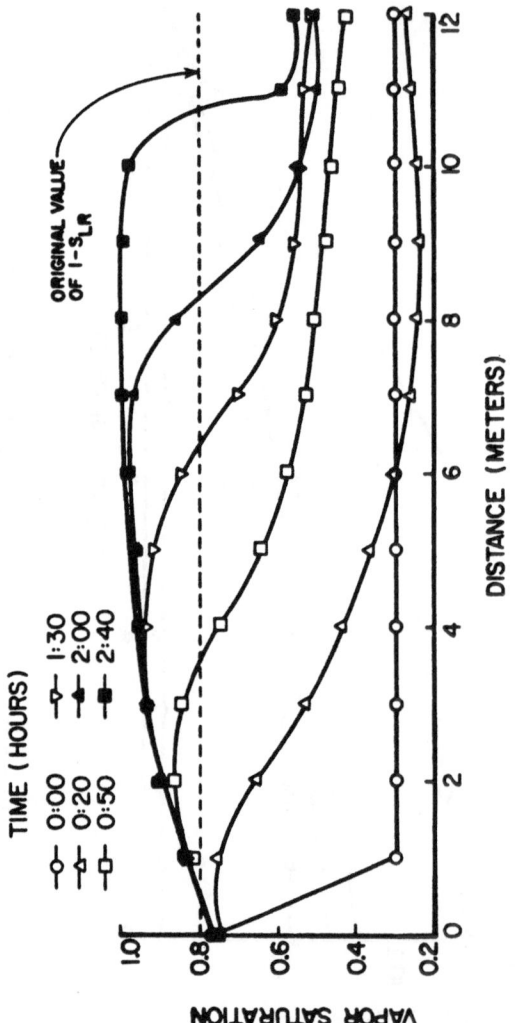

Figure 4-12. Saturation profiles for the vaporizing gas drive.

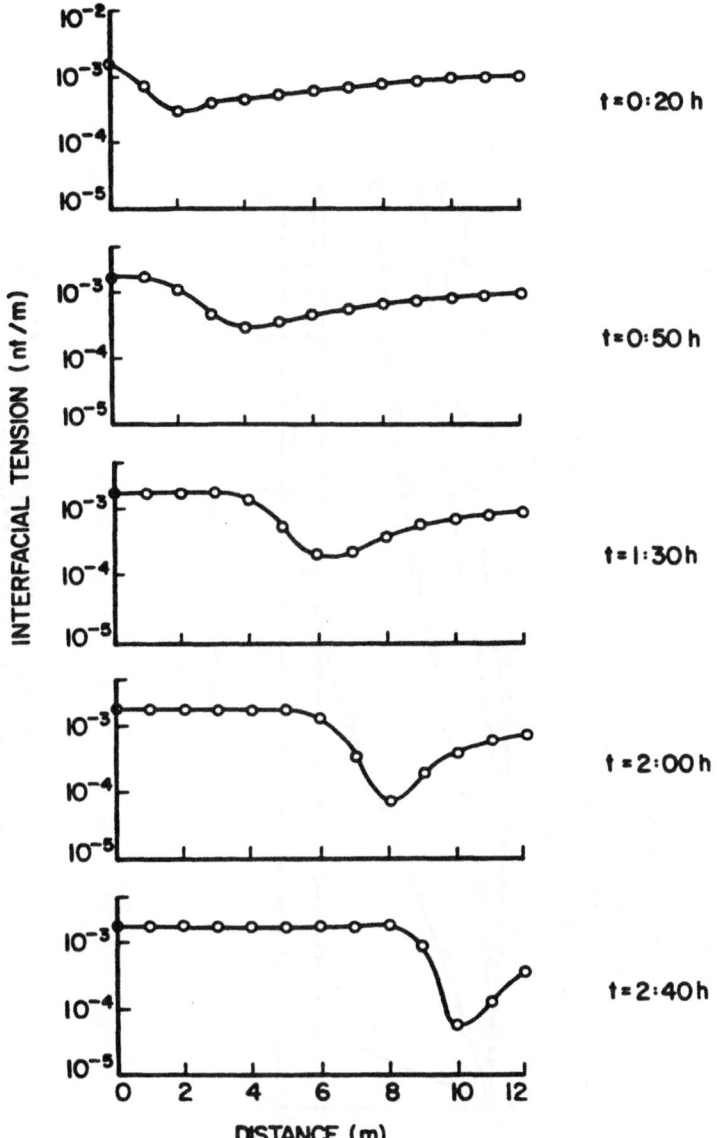

Figure 4-13. Interfacial-tension profile at various times for the vaporizing gas drive.

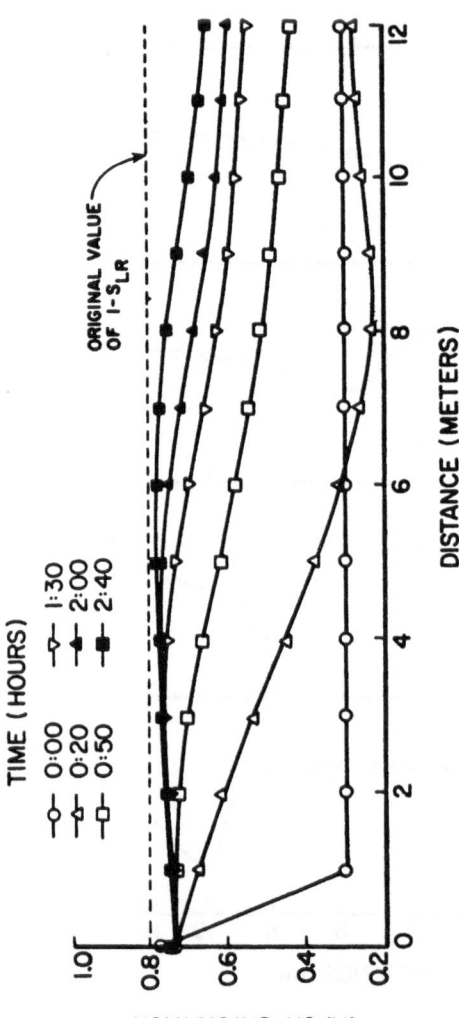

Figure 4-14. Saturation profiles for the vaporizing gas flood, as predicted by orthogonal collocation.

does not accompany any appreciable material balance errors. As in the
Buckley-Leverett problem examined in Chapter Three, orthogonal collocation and upstream collocation deliver numerical solutions that are
qualitatively different in structure. In light of the discussions in
Sections 1.5 and 3.4, the universal adherence to upstream weighting in
finite difference models, and the failure of orthogonal collocation to
reproduce the development of a miscibile transition zone, there are good
grounds here for preferring upstream collocation.

CHAPTER FIVE
CONCLUSIONS

Let us close this investigation with a summary of its main points.
In doing so, we shall briefly discuss the significance of the work as
well as some limitations that may be overcome through further research.

Compositional flows in porous media pose two general problems to the
modeler. One is the task of representing the thermodynamics of fluid
phase behavior. The exchange of molecular species among phases in
contact is an essential feature of such compositional flows as miscible
gas floods, and methods for calculating this exchange have a great deal
to do with the overall effectiveness of numerical simulators. The
second issue confronting modelers is that of solving the partial differential equations governing multicomponent fluid flows. To date this
problem has been the nearly exclusive province of finite-difference
techniques, with finite-element Galerkin methods having made some
inroads. By comparison, finite-element collocation is a newcomer to the
scene, despite its potential computational advantages. The preceding
chapters introduce new approaches for both the thermodynamic calculations and the collocation solution of species transport equations.

The thermodynamic interpolation scheme described in Chapter Two
provides an attractive alternative to the usual equation-of-state calculations of fluid phase behavior. The new approach essentially moves the
difficult solution of vapor-liquid equilibrium constraints from the
coefficient calculations in transport codes to the construction of input
data sets, replacing expensive nonlinear algebraic algorithms by cheap,
reliable geometric ones during execution time. Of course, unlike the
standard equation-of-state methods, the interpolation scheme cannot be

used "off the shelf", as the modeler must construct a separate data set
for each new thermodynamic system. For extended reservoir studies this
disadvantage should be slight compared to gains in computational
efficiency and reliability.

While the particular interpolation scheme constructed in Chapter Two
applies to three-component systems, it has clear extensions to more
complex isothermal systems. The method of plates used in Section 2.3
fits a plane segment of two dimensions through each triple of adjacent
points over a proper triangulation of the two-dimensional composition
space. The resulting assemblage of plane segments then forms the
approximate representation of the Maxwell set, or saturation-pressure
dome, of the system. With N fluid-phase components (or lumped pseudo-
components) one can fit a hyperplane segment of N - 1 dimensions through
each N-tuple of adjacent points in a proper division of the (N-1) -
dimensional composition space into N-simplexes (Oden and Reddy, 1976,
Section 6.6). This interpolation scheme is the logical generalization
of the method of plates to larger numbers of thermodynamic variables.
In the construction of such a scheme it will be necessary to account for
the additional compositional variables in parametrizing the vapor-liquid
tie lines.

Of the results presented in the foregoing chapters, the collocation
techniques perhaps have the broadest potential for application to numerical fluid mechanics. Chapters Three and Four demonstrate the applicability of the method to complicated multiphase flows in porous media,
thus vindicating the method despite early disappointing results
(Sincovec, 1977). Upstream collocation plays a central role in these
applications, especially in nonlinear flows where convective forces
dominate dissipative mechanisms. By collocating convective terms at
points upstream of the usual Gauss points, one can introduce an artifi-

cially dissipative spatial error that forces convergence to physically correct weak solutions in singular-perturbation problems while preserving numerical consistency. This technique should find applications in the simulation of other petroleum reservoir flows and, indeed, in a wide range of convection-dominated problems.

The compositional simulator developed in Chapter Four models the movements of only two fluid phases, with the objective of capturing the essential physics of hydrocarbon flow with interphase mass transfer. To be applicable to natural petroleum reservoirs, a full-scale compositional model should include a transport equation for brine. One common approach to doing this is to treat brine as a separate, aqueous phase that does not exchange molecular species with the hydrocarbon liquid and vapor phases (see, for examples, Coats, 1980, and Nghiem et al., 1981). Thus the behavior of the brine will typically be simpler to model than that of the vapor and liquid phases treated in this study.

The development of collocation-based compositional models for practical engineering use will also require the extension of the formulation given in Chapter Four to more than one spatial dimension. Perhaps the most promising strategy for doing this is to use finite-element bases generated by the tensor products of Hermite cubic bases. There then remains the technical problem of choosing upstream collocation points. In contrast to the one-dimensional case, multidimensional upstream collocation requires the determination of directions that are locally upstream. For porous-media flows this may be accomplished, for example, by examining pressure gradients. The details of such a procedure offer a natural avenue for further inquiry.

In the final analysis the extension of the compositional model to two space dimensions may suffice for many design purposes. Multicomponent flows are expensive to simulate, and compositional models typically make intensive demands on computer storage. As a consequence, fully three-dimensional reservoir models of such flows may require spatial discretizations too coarse to capture the development of miscible transition zones between injection wells and production wells. When this is the case, modelers might be better off using a one- or two-dimensional compositional simulator as a near-field model of the development of miscibility starting at injection wells, resorting to a simpler miscible flood simulator to model the far-field displacements in two or three dimensions.

APPENDIX A

SUMMARY OF MATHEMATICAL NOTATION

1. Sets: membership and specification.

$a \in A$ a is an element of the set A.

$\{a,b,c\}$ The set containing elements a, b, and c.

$\{a \in A | \; P(a) \}$ The set of all elements of A that satisfy the predicate P.

$A \subset B$ The set A is a subset of the set B. In other words, $a \in A$ implies $a \in B$.

$A \cup B$ The union of sets A and B.

$A \cap B$ The intersection of sets A and B.

$A \setminus B$ The set of elements of A that are not elements of B.

2. Ordered pairs, functions, and indexed sets.

$A \times B$ The collection of all ordered pairs (a,b) such that $a \in A$ and $b \in B$.

$f: A \to B$ The function f taking as arguments elements of the set A and yielding values in the set B.

dom(f)   The domain of the function f; in the previous definition, dom(f) = A.

ran(f)   The range of the function f; in the example above, ran(f) is the set of elements of B which are images of elements of A under the function f.

$\{a_i\}_{i=1}^N$ The indexed set $\{a_1, a_2, \ldots, a_N\}$

3. Real numbers.

R   The set of all real numbers.

R^N The set of ordered N-tuples $(r_1, \ldots, r_N)$ with each $r_i \in R$. Euclidean N-space.

sup A   The least upper bound of the subset A of R (provided it exists).

inf A   The greatest lower bound of the subset A of R (provided it exists).

[a,b]   The closed interval containing all real numbers r satisfying $a \leq r \leq b$.

(a,b)   The open interval containing all real numbers r satisfying $a < r < b$.

(a,b], [a,b)   The half-open intervals defined by the conditions $a < r \leq b$, $a \leq r < b$, respectively.

4. Differentiation.

$d_t f$ The total derivative of f with respect to t.

$\partial_t f$ The partial derivative of f with respect to t.

$D_t f$ The material derivative of f.

5. Miscellaneous.

M^T The transpose of the matrix **M**.

trace(T)   The sum of the diagonal elements of the tensor T. The trace of a square matrix is defined similarly.

$C^k(\Omega)$ The space of all functions $f: \Omega \to \mathbf{R}$ such that f and its derivatives through order k are continuous on Ω.

$\|\bullet\|_{L^2}(\Omega)$ The L^2 norm on functions $f: \Omega \to \mathbf{R}$, defined by $(\int_\Omega f^2 dx)^{\frac{1}{2}}$. Abbreviated as $\|\bullet\|_{L^2}$, where Ω is understood from the context.

$L^2(\Omega)$ The space of all functions $f: \Omega \to \mathbf{R}$ such that $\|f\|_{L^2} < \infty$. Two such functions $f, g \in L^2(\Omega)$ are regarded as equivalent if $\|f - g\|_{L^2} = 0$.

$\partial\Omega$ The boundary of the set Ω.

$\|\bullet\|_\infty$ The maximum norm on functions $f: \Omega \to \mathbf{R}$, where Ω is understood from the context. $\|f\|_\infty = \sup_\Omega \{f(x)\}$.

APPENDIX B
THERMODYNAMICS OF RESERVOIR FLUIDS
FROM A GRADIENT-DYNAMIC VIEWPOINT

This appendix outlines a picture of thermodynamics that is logically compatible with the time-varying nature of miscible gas floods but yields classical results useful in the calculation of fluid properties by equilibrium methods. This picture also furnishes a conceptual link between the geometry of thermostatic equilibria and the algebraic descriptions derived by Gibbs (1876 and 1878). The dynamic viewpoint follws that proposed by Gilmore (1981), and the equilibrium results are just the Gibbs conditions for static systems. Thus this appendix contains no essentially new results but establishes a foundation for the equation-of-state methods presented in Chapter Two.

Gradient-dynamic postulate.

For the isothermal fluid system in the reservoir let us identify two distinct sets of thermodynamic variables. The first, C, is the control set containing ordered N-tuples $(c_1, \ldots, c_N) = (\omega_1, \ldots, \omega_{N-1}, V)$ where ω_i is the mole fraction of species i and V is the molar fluid volume, that is, the reciprocal of molar fluid density ρ. The second set, X, is the state space containing N-tuples $(s_1, \ldots, s_N) = (\eta_1, \ldots, \eta_{N-1}, -p)$ where η_i is the (modified) chemical potential of species i and p is the pressure. The variables p and V are positive, and the mole fractions must lie in $[0,1]$.

We shall postulate that the thermodynamic system obeys a gradient-dynamic law. Specifically, let us assume that there exists an analytic function $A : X \times C \to R$, called the potential and having dimensions $[ML^2/(T^2 mol)] = [energy/mole]$, such that

$$d_t \underset{\sim}{s} = - \nabla_s \ A \ (\underset{\sim}{s}, \underset{\sim}{c})$$

$$(B-1)$$

where $\nabla_s \ A$ is the vector with elements $\partial \ A \ / \partial s_i$. This postulate is almost surely a simplification of nature; however, it is perhaps a plausible approximation for systems not far from stasis. An equation of the form (B-1) is in a sense the simplest non-static form of dynamics one can assume for the thermodynamic system (Gilmore, 1981, Chapter 1).

The equilibrium set M_A of the system (B-1) is the set of points in $X \times C$ where $d_t s_i = 0$, $i = 1, \ldots, N$:

$$M_A = \{(\underset{\sim}{s}, \underset{\sim}{c}) \ \epsilon \ X \times C \ | \ \nabla_s \ A \ (\underset{\sim}{s}, \underset{\sim}{c}) = 0\}$$

$$(B-2)$$

A point $(\underset{\sim}{s}, \underset{\sim}{c}) \ \epsilon \ M_A$ is a stable equilibrium if A has a local minimum at $(\underset{\sim}{s}, \underset{\sim}{c})$ or, equivalently, if the determinant of $\text{Hess}_s[\ A \ (\underset{\sim}{s}, \underset{\sim}{c})]$ is positive, where $\text{Hess}_s(\ A \ )$ denotes the Hessian matrix $\partial^2 A \ / \partial s_i \partial s_j$. Let us call $(\underset{\sim}{s}, \underset{\sim}{c}) \ \epsilon \ M_A$ a thermostatic equilibrium if A has a global minimum there.

These notions of equilibrium correspond to those proposed by Gibbs. The thermodynamic system governed by (B-1) exhibits qualitative changes in the behavior of stable equilibria at points of C where local minima in the potential A disappear, that is, where $\det[\text{Hess}_s(\ A \ )] = 0$. The set

$$K_S = \{(\underset{\sim}{s}, \underset{\sim}{c}) \ \epsilon \ M_A \ | \ \det[\text{Hess}_s \ A \ (\underset{\sim}{s}, \underset{\sim}{c})] = 0\}$$

$$(B-3)$$

of such points is the spinodal set of A. In classical thermostatics, K_S is the boundary of metastability; in Gibbs' language, K_S is the boundary of stability against "continuous changes of phase", by which he means perturbations not involving the formation of new phases (Gibbs, 1876 and 1878).

The thermodynamic system exhibits qualitative changes in the behavior of thermostatic equilibria at points of C where the global minimum in A shifts from one local minimum to another, so that global equiminima occur. The set

$$K_M = \{(\underset{\sim}{s},\underset{\sim}{c}) \in M_A \mid (\underset{\sim}{s},\underset{\sim}{c}) \text{ is a global minimum}$$
$$\text{in } A \text{ and } A(\underset{\sim}{s},\underset{\sim}{c}) = A(\underset{\sim}{s}',\underset{\sim}{c}') \text{ for some}$$
$$(\underset{\sim}{s}',\underset{\sim}{c}') \in M_A \text{ with } \underset{\sim}{s}' \neq \underset{\sim}{s}\}$$

$$(B-4)$$

corresponding to shifts in global minima is the Maxwell set of A. Classically, K_M is the boundary between different equilibrium phase regimes: on crossing K_M a system in thermostatic equilibrium will manifest new phases or the disappearance of previously existing phases. In Gibbs' terms, K_M is the boundary of stability against "discontinuous changes of phase", that is, perturbations allowing the formation of new phases (Gibbs, 1876 and 1878).

As Gilmore (1981, Chapter 10) shows, it is possible to construct a formal identification between the equilibria of the system (B-1) and Gibbs' thermostatics. By the implicit function theorem, the conditions $\nabla_s A = 0$ defining M_A implicitly define an equation of state $\underset{\sim}{s} = \underset{\sim}{s}(\underset{\sim}{c})$ at all points $(\underset{\sim}{s},\underset{\sim}{c}) \in M_A$ for which $\text{Hess}_s(A)$ is nonsingular. Let $I_c: M_A \rightarrow C$ be the projection of such points into the control space, so that $I_c(\underset{\sim}{s}(\underset{\sim}{c}),\underset{\sim}{c}) = \underset{\sim}{c}$. The function $\hat{A}: I_c(M_A) \rightarrow R$ defined by

$$\hat{A}(\underset{\sim}{c}) = A\ (\underset{\sim}{s}(\underset{\sim}{c}),\underset{\sim}{c})$$

$$(B-5)$$

is the molar Helmholtz free energy.

To complete the correspondence with Gibbs' thermodynamics, identify

$$\underset{\sim}{s} = \nabla_c \hat{A}(\underset{\sim}{c})$$

$$(B-6a)$$

In particular, the pressure is given by

$$- p = \partial_V \hat{A}$$

$$(B-6b)$$

and the modified chemical potentials are

$$\eta_i = \partial_{\omega_i} \hat{A}, \quad i = 1,\ldots,N-1$$

$$(B-6c)$$

It is worth noting that the η_i differ from the usual chemical potentials μ_i, which are derivatives of A with respect to mole numbers. At stable equilibrium the two stand in the relation $\eta_i = \mu_i - \mu_N$, $i = 1,\ldots,N-1$ (Reid and Beegle, 1971).

Local thermostatic equilibrium.

Given the thermodynamic formalism outlined above, it is meaningful to ask of a given system whether it evolves with very small departures from thermostatic equilibrium. Without examining the detailed experimental data needed to answer this question for actual systems, let us briefly discuss some considerations that lend coherence to the notion of local thermostatic equilibrium in miscible gas floods. The gist of the argument is that the time scales characteristic of changes governed by the transport equations may be much longer than those characteristic of the approach to equilibrium governed by (B-1). Hence, to a very good approximation the flowing fluids behave as if their thermodynamic variables instantaneously attained values corresponding to local thermostatic equilibrium.

The transport equations developed in Chapter One are material balances:

$$\partial_t(\rho\omega_i) + \nabla\cdot(\rho\omega_i\underset{\sim}{v}_i) = 0, \qquad i = 1,\ldots,N$$

$$(B-7)$$

where

$$\underset{\sim}{v}_i = \phi(S_V\rho^V\omega_i^V\underset{\sim}{v}^V + S_L\rho^L\omega_i^L\underset{\sim}{v}^L)/(\rho\omega_i)$$

$$(B-8)$$

is the mean velocity of species i in the fluids. If we define the mean fluid velocity as $\underset{\sim}{v} = \Sigma_{i=1}^{N} \rho\, \omega_i\, \underset{\sim}{v}_i$, sum equation (B-7) over i, and denote $D_t = \partial_t + \underset{\sim}{v}\cdot\nabla$, we find

$$D_t V = V^2 \, \nabla\cdot(\underset{\sim}{v}/V)$$

$$(B-9)$$

where V is the molar volume. In light of this equation and the definition $u_i = v_i - v$, equation (B-7) becomes

$$D_t \omega_i = - V \nabla \cdot (\omega_i \underset{\sim}{u}_i / V), \qquad i = 1, \ldots, N-1$$

$$(B-10)$$

Therefore the transport equations are equivalent to evolution equations
for the control variables $(\omega_i, \ldots, \omega_{N-1}, V) = c$ of the thermodynamic
system. Conceptually,

$$D_t \underset{\sim}{c} = F(\underset{\sim}{c}, \ldots)$$

$$(B-11)$$

where the ellipsis allows for dependence on other quantities, including
spatial gradients of control variables.

For systems in which $D_t \underset{\sim}{c} \neq 0$ there is no reason to believe that $d_t \underset{\sim}{s} = 0$ exactly. Suppose, though, that the potential $A(\underset{\sim}{s}, \underset{\sim}{c})$ and its
equilibria change in response to (B-11) at a rate comparable to $(1/T_c) = \max_i \{(1/c_i^0) D_t c_i\}$, where c_i^0 is some reference value characteristic of
the system. In nature thermodynamic systems exhibit time scales charac-
teristic of the relaxation to equilibrium, governed in this case by
(B-1). For simple systems such relaxation may obey an exponential decay
law, while for multispecies systems the behavior can be more complex
(see Bird et al., 1960, Chapter 21). Let τ_i be a characteristic time
for s_i to relax to thermostatic equilibrium, and call $\tau_{max} = \max_i \{\tau_i\}$.
Then the departure of the system from thermostatic equilibrium will be
small provided $T_c \gg \tau_{max}$ (Gilmore, 1981, Chapter 8). In this case a
material point evolves essentially as if it were confined to the
equilibrium set M_A, bifurcating when it intersects the Maxwell set K_M.
In all that follows let us assume that this condition holds.

Constraints of multiphase equilibria.

For single-phase fluids the constraints of local thermostatic
equilibrium amount to the requirement that A assume its unique global
minimum for the given values $\underset{\sim}{c}$, and an equilibrium equation of state
supplies all of the necessary information. In compositional flows,
however, substantial regions of the reservoir may contain more than one
fluid phase, each of which corresponds to a distinct global equiminimum
of A . For such regions the coexisting fluid phases must satisfy not
only the equation of state but also the criterion that they lie on the
Maxwell set K_M. This criterion is equivalent to a well-established set
of algebraic constraints whose derivation dates from Gibbs' work. We
shall reproduce the arguments here in language that is consistent with
the gradient-dynamic postulate. The clear explications of Gibbs (1876
and 1878), Münster (1970, Chapter 7), and Gilmore (1981, Chapter 5)
guide the discussions that follow.

If, at a particular time and place in the reservoir, a vapor and a
liquid stand in local thermostatic equilibrium, then there must exist
two global equiminima in A , say $(\underset{\sim}{s}^V,\underset{\sim}{c}^V)$, $(\underset{\sim}{s}^L,\underset{\sim}{c}^L) \in K_M$, with $\underset{\sim}{s}^V \neq \underset{\sim}{s}^L$.
Consider the response of these minima to changes δc_i in the control
parameters. By Taylor's theorem, since $\partial A / \partial s_i = 0$ at $(\underset{\sim}{s}^\alpha,\underset{\sim}{c}^\alpha)$ for $i =$
$1,\ldots,N$ and $\alpha = V$ or L,

$$A (\underset{\sim}{s}^\alpha + \delta\underset{\sim}{s}^\alpha, \underset{\sim}{c}^\alpha + \delta\underset{\sim}{c}) = A^\alpha + \nabla_c A \cdot \delta\underset{\sim}{c} + O((\delta\underset{\sim}{s}^\alpha + \delta\underset{\sim}{c})^2)$$

$$(B-12)$$

where the superscript α indicates evaluation at $(\underset{\sim}{s}^\alpha,\underset{\sim}{c}^\alpha)$, $\alpha = V$ or L.
Thus, to first order in the thermodynamic variables, the response of the
minima $A (\underset{\sim}{s}^\alpha,\underset{\sim}{c}^\alpha)$ to small changes in the control parameters is

$$\delta \ A^{\alpha} = A \ (\underset{\sim}{s}^{\alpha} + \delta \underset{\sim}{s}^{\alpha}, \ \underset{\sim}{c}^{\alpha} + \delta \underset{\sim}{c}) - A^{\ \alpha} = \nabla_{c} \ A \ (\underset{\sim}{s}^{\alpha}, \underset{\sim}{c}^{\alpha}) \cdot \delta \underset{\sim}{c}$$

$$(B-13)$$

where α = V or L.

Using equation (B-13) it is possible to derive conditions defining the Maxwell set, K_{M}. For a change $\delta \underset{\sim}{c}$ applied to points $(\underset{\sim}{s}^{V}, \underset{\sim}{c}^{V})$, $(\underset{\sim}{s}^{L}, \underset{\sim}{c}^{L})$ $\epsilon \ K_{M}$ to yield perturbed states that also lie in K_{M}, it is necessary that $\delta \underset{\sim}{c}$ force equal changes in the equal (minimum) values of A, so that $\delta \ A^{V} = \delta \ A^{L}$. In symbols,

$$[\nabla_{c} \ A(\underset{\sim}{s}^{V}, \underset{\sim}{c}^{V}) - \nabla_{c} \ A(\underset{\sim}{s}^{L}, \underset{\sim}{c}^{L})] \cdot \delta \underset{\sim}{c} = 0$$

$$(B-14)$$

Gilmore (1981, Chapter 5) calls this a generalized Clausius-Clapeyron equation.

The demand that $(\underset{\sim}{s}^{V}, \underset{\sim}{c}^{V})$, $(\underset{\sim}{s}^{L}, \underset{\sim}{c}^{L})$ and the perturbed points $(\underset{\sim}{s}^{V} + \delta \underset{\sim}{s}^{V}$, $\underset{\sim}{c}^{V} + \delta \underset{\sim}{c})$, $(\underset{\sim}{s}^{L} + \delta \underset{\sim}{s}^{L}, \ \underset{\sim}{c}^{L} + \delta \underset{\sim}{c})$ all lie in $K_{M} \subset M_{A}$ allows us to rewrite the potential derivatives in terms of the Helmholtz free energy, so by equations (B-6),

$$\nabla_{c} \ A(\underset{\sim}{s}^{\alpha}, \underset{\sim}{c}^{\alpha}) = \nabla_{c} A(\underset{\sim}{c}^{\alpha}) = \underset{\sim}{s}^{\alpha}$$

$$(B-15)$$

where α = V or L. Since δc is an arbitrary small perturbation, substituting (B-15) into (B-14) for each phase gives

$$s_{i}^{V} = s_{i}^{L}, \quad i = 1, \ldots, N$$

$$(B-16a)$$

In particular,

$$\eta_i^V = \eta_i^L, \quad i = 1,\ldots,N-1$$

(B-16b)

$$P_V = P_L$$

(B-16c)

These are precisely Gibbs' conditions.

Critical point criteria.

Gibbs (1876 and 1878, pp. 129-133) deduces two algebraic criteria for critical points of multicomponent systems. The first condition for a point $(\underset{\sim}{s}(\underset{\sim}{c}),\underset{\sim}{c}) \in M_A$ to be a critical point is that it lie at the limit of thermostatic equilibria for which two phases coexist, that is, at the point where equiminima merge. This requires that the critical point be an inflection point of the Helmholtz free energy:

$$\det (\partial_{c_i} \partial_{c_j} \hat{A}) = 0$$

(B-17a)

or, in expanded form,

$$\det \begin{bmatrix} \partial_{\omega_1}^2 \hat{A} & \cdots & \partial_{\omega_1} \partial_V \hat{A} \\ & & \\ \cdot & & \\ \cdot & & \\ \cdot & & \\ \partial_V \partial_{\omega_1} \hat{A} & \cdots & \partial_V^2 \hat{A} \end{bmatrix} = U = 0$$

(B-17b)

The second criterion is that $(s(c),c)$ be a limit of points satisfying (B-17) for which isothermal variations in the control parameters at equilibrium can produce unstable phases. Since the system is confined to the equilibrium set M_A, the condition $\det(\mathrm{Hess}_s\ A) < 0$ for instability reduces through the chain rule to the condition $U < 0$. Thus the second criterion for critical points implies that the determinant U does not become negative under isothermal perturbations in control parameters, so long as the perturbations leave the system in M_A. Under such a perturbation $\delta\underset{\sim}{c}$,

$$U(\underset{\sim}{c} + \delta\underset{\sim}{c}) = U(\underset{\sim}{c}) + \nabla_c U(\underset{\sim}{c})\cdot\delta\underset{\sim}{c} + O(\delta\underset{\sim}{c}^2)$$

$$(B\text{-}18)$$

So, to have $U(\underset{\sim}{c} + \delta\underset{\sim}{c}) - U(\underset{\sim}{c}) \geq 0$ for arbitrary isothermal variations confined to equilibrium, we must have

$$\nabla_c U(\underset{\sim}{c})\cdot\delta\underset{\sim}{c} = 0$$

$$(B\text{-}19)$$

Now for equations (B-17) and (B-19) to have a solution, they must be consistent. In particular the null space of the matrix in (B-17) must remain invariant when any of the matrix rows is replaced by the vector $(\partial U/\partial c_1, \ldots, \partial U/\partial c_N)$. For this it is necessary (Münster, 1970, Chapter 7) that

$$\det \begin{bmatrix} \partial_{\omega_1} U & \cdots & \partial_V U \\ \partial_{\omega_2}\partial_{\omega_1}\hat{A} & \cdots & \partial_{\omega_2}\partial_V\hat{A} \\ \cdot & & \\ \cdot & & \\ \cdot & & \\ \partial_V\partial_{\omega_1}\hat{A} & \cdots & \partial_V^2\hat{A} \end{bmatrix} = 0$$

$$(B\text{-}20)$$

The critical conditions (B-17) and (B-20) assume a form involving lower-order determinants if we apply the Legendre transformation $\hat{A}(\omega_1,\ldots,\omega_{N-1}, V) - pV = \hat{G}(\omega_1,\ldots,\omega_{N-1}, p)$ defining the Gibbs free energy (Reid and Beegle, 1977):

$$\det \begin{bmatrix} \partial^2_{\omega_1} \hat{G} & \cdots & \partial_{\omega_1} \partial_{\omega_{N-1}} \hat{G} \\ \cdot & & \\ \cdot & & \\ \cdot & & \\ \partial_{\omega_{N-1}} \partial_{\omega_1} \hat{G} & \cdots & \partial^2_{\omega_{N-1}} \hat{G} \end{bmatrix} = U = 0 \qquad \text{(B-21)}$$

$$\det \begin{bmatrix} \partial_{\omega_1} U & \cdots & \partial_{\omega_{N-1}} U \\ \partial_{\omega_2} \partial_{\omega_2} \hat{G} & \cdots & \partial_{\omega_2} \partial_{\omega_{N-1}} \hat{G} \\ \cdot & & \\ \cdot & & \\ \cdot & & \\ \partial_{\omega_{N-1}} \partial_{\omega_1} \hat{G} & \cdots & \partial^2_{\omega_{N-1}} \hat{G} \end{bmatrix} = 0 \qquad \text{(B-22)}$$

APPENDIX C
THE CORRESPONDENCE BETWEEN ORTHOGONAL
COLLOCATION AND GALERKIN'S METHOD

   This appendix reviews the algebraic correspondence between orthogonal collocation and an approximate Galerkin scheme in one spatial dimension. Various investigators, including Douglas and Dupont (1973), Prenter (1975, Section 8.8), and Botha and Pinder (1983, Section 4.4) have developed this argument as a means of analyzing collocation. Let us examine, as a paradigm, the constant-coefficient convection-dispersion equation of Section 3.2:

$$\partial_t \omega - D_i \, \partial_x^2 \omega + v \, \partial_x \omega = 0$$

$$(C-1)$$

on a spatial domain $\Omega = [0, x_{max}]$. To simplify the exposition, let us consider the method of lines obtained by discretizing only the spatial dimension.

   Given a uniform partition Δ_M of Ω with M nodes and mesh Δx, the trial function for $\omega(x,t)$ in the Hermite space $H_3(\Delta_M)$ has the form

$$\hat{\omega}(x,t) = \omega_\partial(x,t) + \sum_{m=1}^{K} W_m(t) \, H_m(x)$$

$$(C-2)$$

Here, $K = 2M - 2$; $\omega_\partial(x,t) \in H_3(\Delta_M)$ satisfies the boundary conditions, and the interior basis functions satisfy

$$H_m(0) = H_m(x_{max}) = 0, \quad m = 1,\ldots,2M-2$$

$$(C-3)$$

Equation (C-2) is just a compact way of writing the Hermite cubic trial function without using double subscripts to distinguish the basis functions. For the boundary conditions (3.2-4), for example, the relationship between the double-subscript notation and the compact notation is as follows:

$$H_{j,\ell}(x) = H_{2\ell-2+j}(x), \quad j = 0,1, \quad \ell = 1,\ldots,M$$

$$(C-4)$$

Substituting (C-2) into the left side of (C-1) yields a residual

$$R(x,t) = \partial_t \hat{\omega} - D_i \, \partial_x^2 \hat{\omega} + v \, \partial_x \hat{\omega}$$

$$(C-5)$$

In terms of this quantity, collocation demands

$$R(x_k,t) = 0, \quad k = 1,\ldots,2M-2$$

$$(C-6)$$

while the Galerkin method requires

$$\int_\Omega R(x,t) \, H_n(x) \, dx = 0, \quad n = 1,\ldots,2M-2$$

$$(C-7)$$

Consider an approximation to (C-7) in which the integrals are replaced by two-point Gauss quadratures on each element $[\bar{x}_\ell, \bar{x}_{\ell+1}]$ of the partition Δ_M:

$$\int_\Omega R(x,t)\, H_n(x)\, dx = \sum_{m=1}^{K} R(x_k,t)\, H_n(x_k) + O(\Delta x^5)$$

$$(C-8)$$

Here the values x_k are Gauss points, given by roots of the quadratic Legendre polynomials native to each element. These roots $x_\ell + \frac{1}{2} \Delta x \pm \Delta x \sqrt{3}$ are precisely the orthogonal collocation points. If we neglect the error $O(\Delta x^5)$, then substituting (C-8) into (C-7) renders the approximate Galerkin scheme

$$\sum_{k=1}^{K} \sum_{m=1}^{K} [d_t W_m(t)\, H_m(x_k) - D_i\, W_m(t)\, d_x^2 H_m(x_k)$$

$$+ v\, W_m(t)\, d_x H_m(x_k)]\, H_n(x_k) = 0,$$

$$(C-9)$$

for $n = 1, \ldots, K$.

We can unravel these equations by defining arrays **B**, **E**, and **W**· as follows:

$$B_{n,k} = H_n(x_k)$$

$$(C-10a)$$

$$E_{k,m} = D_i\, d_x^2 H_m(x_k) + v\, d_x H_m(x_k)$$

$$(C-10b)$$

$$W\bullet_k = \sum_{m=1}^{K} d_t W_m(t) \, H_m(x_k),$$

$$(C-10c)$$

where the indices k, m, and n range over 1,...,K. In terms of these and the vector $W = (W_1,...,W_{2M-2})^T$, the approximate Galerkin scheme (C-9) is

$$B \, W\bullet = - \, B \, E \, W$$

$$(C-11)$$

But orthogonal collocation is simply

$$W\bullet = - \, E \, W$$

$$(C-12)$$

Douglas and Dupont (1973) prove that the matrix **B** is invertible, so (C-11) and (C-12) are equivalent.

BIBLIOGRAPHY

Abbott, M. M. (1979), "Cubic equations of state: an interpretive
   review." In: *Equations of State in Engineering and Research*, ed. by
   K. C. Chao and R. L. Robinson. Washington, D.C.: American Chemical
   Society.

Allen, M. B. (to appear), "How upstream collocation works." *Int. J.
   Num. Meth. Engrg.*

Allen, M. B., and Pinder, G. F. (1982), "The convergence of upstream
   collocation in the Buckley-Leverett problem." SPE 10978, presented at
   the 57th Annual Fall Technical Conference and Exhibition of the
   Society of Petroleum Engineers of AIME, New Orleans, September 26-29,
   1982.

Allen, M. B., and Pinder, G. F. (1983), "Collocation simulation of
   multiphase porous-medium flow." *Soc. Pet. Eng. J.* (February),
   135-142.

Amaefule, J. O., and Handy, L. L. (1982), "The effect of interfacial
   tensions on relative oil/water permeabilities of consolidated porous
   media." *Soc. Pet. Eng. J.* (June), 371-381.

Ames, W. F. (1977), *Numerical Methods for Partial Differential
   Equations*. New York: Academic Press.

Aronofsky, J. S., and Jenkins, R. (1951), "Unsteady flow of gas through
   porous media, one-dimensional case." In *Proceedings of the First
   National Congress of Applied Mechanics*, held in Chicago, 763-771.

Asselineau, L., Bogdanic, G., and Vidal, J. (1979), "A versatile algorithm for calculating vapor-liquid equilibria." *Fluid Phase Equilibria* **3**, 273-290.

Atkin, R. J., and Craine, R. E. (1976), "Continuum theories of mixtures: basic theory and historical development." *Quart. Jour. Appl. Math.* **29**: 2, 209-244.

Aziz, K., and Settari, A. (1979), *Petroleum Reservoir Simulation*. London: Applied Science.

Baker, L. E., and Luks, K. D. (1980), "Critical point and saturation pressure calculations for multicomponent systems." *Soc. Pet. Eng. J.* (February), 75-80.

Banjia, V. K., Bennett, C., Reynolds, A., Raghavan, R., and Thomas, G. (1978), "Alternating direction collocation methods for simulating reservoir performance." SPE 7414, presented at the 53th Annual Fall Technical Conference and Exhibition of the Society of Petroleum Engineers of AIME, Dallas, October 1-3, 1978.

Bardon, C., and Longeron, D. G. (1980), "Influence of very low interfacial tensions on relative permeability." *Soc. Pet. Eng. J.* (October), 391-401.

Bird, R. B., Stewart, W. E., and Lightfoot, E. N. (1960), *Transport Phenomena*. New York: John Wiley and Sons.

Birkhoff, G. (1983), "Numerical fluid dynamics." *SIAM Review* **25**: 11, 1-34.

de Boor, R. D., and Swartz, B. (1973), "Collocation at Gaussian points." *SIAM Jour. Numer. Anal.* **10:** 4, 582-606.

Boris, J. P., and Book, D. L. (1976), "Solution of continuity equations by the method of flux-corrected transport." In: *Methods of Computational Physics* **16**, ed. by B. Alder, S. Fernbach, and M. Rotenberg. New York: Academic Press, 85-129.

Botha, J. W., and Pinder, G. F. (1983), *Fundamental Concepts in the Numerical Solution of Partial Differential Equations.* New York: John Wiley and Sons (in press).

Bowen, R. M. (1980), "Incompressible porous media models by the use of the theory of mixtures." *Int. J. Engrg. Sci.* **18**, 787-800.

Bowen, R. M. (1982), "Compressible porous media models by use of the theory of mixtures." *Int. J. Engrg. Sci.* **20:** 6, 697-735.

Buckley, S. E., and Leverett, M. C. (1942), "Mechanism of fluid displacement in sands." *Trans. AIME* **146**, 107-116.

Carbonell, R. G., and Whitaker, S. (1982), "Heat and mass transport in porous media." Presented at the NATO Advanced Study Institute on Mechanics of Fluids in Porous Media, Newark, Delaware, July 18-27, 1982.

Cavendish, J. C., Price, H. S., and Varga, R. S. (1971), "Galerkin methods for numerical solution of boundary value problems." *Soc. Pet. Eng. J.* (December), 374-388.

Chase, C. A. (1979), "Variational simulation with numerical decoupling and local mesh refinement." SPE 7680, presented at the Fifth SPE Symposium on Reservoir Simulation, Denver, January 31 - February 2, 1979.

Chaudhari, N. M. (1973), "A numerical solution with second-order accuracy for multi-component compressible stable miscible flow." *Soc. Pet. Eng. J.* (February), 84-92.

Chavent, G., and Salzano, G. (1982), "A finite-element method for 1-D waterflooding problem with gravity." *Jour. Comput. Phys.* **45**, 307-344.

Chawla, T. C., Leaf, G., Chen, W. L., and Grolmes, M. A. (1975), "The application of the collocation method using Hermite cubic splines to nonlinear transient one-dimensional heat conduction problems." *Trans. ASME, Jour. Heat Transfer* (November), 562-568.

Chorin, A. J., and Marsden, J. E. (1979), *A Mathematical Introduction to Fluid Mechanics.* New York: Springer-Verlag.

Christie, I., and Mitchell, A. R. (1978), "Upwinding of high-order Galerkin methods in conduction-convection problems." *Int. J. Num. Meth. Engrg.* **12**, 1764-1771.

Coats, K. (1980), "An equation-of-state compositional model." *Soc. Pet. Eng. J.* (October), 363-376.

Collins, R. E. (1961), *Flow of Fluids through Porous Materials.* Tulsa: Petroleum Publishing Co.

Culham, W. E., and Varga, R. S. (1971), "Numerical methods for time-dependent, nonlinear boundary value problems." *Soc. Pet. Eng. J.* (December), 374-388.

Dahlquist, G., and Bjork, A. (1974), *Numerical Methods.* Englewood Cliffs, New Jersey: Prentice-Hall.

Dalen, V. (1979), "Simplified finite-element methods for reservoir flow problems." *Soc. Pet. Eng. J.* (October), 333-343.

Davis, H. T., and Scriven, L. E. (1980), "The origins of low interfacial tensions for enhanced oil recovery." SPE 9278, presented at the 55th Annual Fall Technical Conference and Exhibition of the Society of Petroleum Engineers of AIME, Dallas, September 21-24, 1980.

Doscher, T. M. (1980), "Enhanced recovery of crude oil." *American Scientist* **69**, 193-199.

Douglas, J., and Dupont, T. (1973), "A finite-element collocation method for quasilinear parabolic equations." *Math. Comp.* **27**: 121, 17-28.

Douglas, J., Kendall, R. P, and Wheeler, M. F. (1979), "Long wave regularization of one-dimensional, two-phase immiscible flow in porous media." In *Finite Element Methods for Convection Dominated Flows*, ed. by T. J. R. Hughes. New York: American Society of Mechanical Engineers, 201-211.

Eringen, A. C., and Ingram, J. D. (1965), "A continuum theory of chemically reacting media -- I." *Int. J. Engrg. Sci.* **3**, 197-212.

Fried, J. J. (1975), *Groundwater Pollution.* Amsterdam: Elsevier Scientific Pub. Co.

Fried, J. J., and Combarnous, M. A. (1971), "Dispersion in porous media." In: *Advances in Hydroscience* **7**, ed. by V. T. Chow. New York: Academic Press, 169-281.

Fussell, L. T., and Fussell, D. D. (1979), "An iterative technique for compositional reservoir models." *Soc. Pet. Eng. J.* (August), 211-220.

Fussell, D. D., and Yanosik, J. L. (1978), "An iterative sequence for phase-equilibria calculations incorporating the Redlich-Kwong equation of state." *Soc. Pet. Eng. J.* (June), 173-182.

Gibbs, J. W. (1876 and 1878), "On the equilibrium of heterogeneous substances." *Trans. Conn. Acad.* **3**, 108-248 and 343-524. Page citations refer to *The Collected Works of J. W. Gibbs, Volume I: Thermodynamics*. New York: Longmans, Green, and Co., 1928.

Gilmore, R. (1981), *Catastrophe Theory for Scientists and Engineers*. New York: Wiley Interscience.

Gray, W. G., and Pinder, G. F. (1976), "An analysis of the numerical solution of the transport equation." *Water Res. Research* **12**: 3, 547-555.

Greenkorn, R. A., and Kessler, D. P. (1969), "Dispersion in heterogeneous nonuniform anisotropic porous media." *Ind. Eng. Chem.* **61**: 9, 33-49.

Gresho, P. M., and Lee, R. L. (1979), "Don't suppress the wiggles -- they're telling you something." In *Finite Element Methods for Convection Dominated Flows*, ed. by T. J. R. Hughes. New York: American Society of Mechanical Engineers, 37-61.

Gundersen, T. (1982), "Numerical aspects of the implementation of cubic equations of state in flash calculation routines." *Computers in Chem. Engrg.* **6**: 3, 245-255.

Gubbins, K. E., and Haille, J. M. (1977), "Molecular theories of inter-facial tension." In: *Improved Oil Recovery by Surfactant and Polymer Flooding*, ed. by D. O. Shah and R. S. Schechter. New York: Academic Press, 119-159.

Hand, D. B. (1930), "Dineric distribution." *Jour. Phys. Chem.* **34**, 1961-2000.

Hassanizadeh, M., and Gray, W. G. (1980), "General conservation equations for multi-phase systems: 3. Constitutive theory for porous media flow." *Adv. Water Resources* **3**, 25-40.

Heidemann, R. A., and Khalil, A. M. (1980), "The calculation of critical points." *AIChE Jour.* **26**: 2, 137-144.

Heinrich, J. C., and Zienkiewicz, O. C. (1977), "Quadratic finite element schemes for two-dimensional convective-transport problems." *Int. J. Num. Meth. Eng.* **11**, 1831-1844.

Helfferich, F. G. (1981), "General theory of multicomponent, multiphase displacement in porous media." *Soc. Pet. Eng. J.* (February), 51-62.

Helfferich, F. G. (1982), "Generalized Welge construction for two-phase flow in porous media in system with limited miscibility." SPE 9730, presented at the 57th Annual Fall Technical Conference and Exhibition of the Society of Petroleum Engineers of AIME, New Orleans, September 26-29, 1982.

Holm, L. W. (1976), "Status of CO_2 and hydrocarbon miscible oil recovery methods." *Jour. Pet. Tech.* (January), 75-83.

Holm, L. W. (1982), "CO_2 flooding: its time has come." *Jour. Pet. Tech.* (December), 2739-2745.

Hughes, T. J. R. (1978), "A simple scheme for developing upwind finite elements." *Int. J. Num. Meth. Engrg.* **12**, 1359-1365.

Hughes, T. J. R., and Brooks, A. (1979), "A multi-dimensional upwind scheme with no crosswind diffusion." In *Finite Element Methods for Convection Dominated Problems*, ed. by T. J. R. Hughes. New York: American Society of Mechanical Engineers, 19-35.

Hutchinson, C. A., and Braun, P. H. (1961), "Phase relations of miscible displacement in oil recovery." *AIChE Jour.* **7**: 1, 64-72.

Huyakorn, P. (1977), "Solution of steady-state, convective transport equation using an upwind finite element scheme." *Appl. Math. Modelling* **1**, 187-195.

Huyakorn, P., and Pinder, G. F. (to appear), *Computational Methods in Subsurface Flow*. New York: Academic Press.

Ingram, J. D., and Eringen, A. C. (1967), "A continuum theory of chemically reacting media -- II. Constitutive equations of reacting fluid mixtures." *Int. J. Engrg. Sci.* **5**, 289-322.

Isaacson, E. L. (1981), "Global solution of a Riemann problem for a non-strictly hyperbolic system of conservation laws arising in enhanced oil recovery." Preprint, Department of Mathematical Physics, Rockefeller University, New York.

Israel, R. B. (1979), *Convexity in the Theory of Lattice Gases*, with Introduction by A. S. Wightman. Princeton, New Jersey: Princeton University Press.

Jensen, O. K., and Finlayson, B. A. (1980), "Oscillation limits for weighted residual methods applied to convective diffusion equations." *Int. J. Num. Meth. Engrg.* **15**, 1681-1689.

John, F. (1981), *Partial Differential Equations*, 4th ed. New York: Springer-Verlag.

Kao, J. (1978), "An algorithm for calculating vapour-liquid equilibrium." SPE 7605, presented at the 53rd Annual Fall Technical Conference and Exhibition of the Society of Petroleum Engineers of AIME, Houston, October 1-3, 1978.

Kazemi, H., Vestal, C. R., and Shank, G. D. (1978), "An efficient multicomponent numerical simulator." *Soc. Pet. Eng. J.* (October), 355-368.

Lake, L. W., Pope, G. A., Carey, G. F., and Sepehrnoori, K. (1981), "Isothermal, multiphase, multicomponent fluid-flow in premeable media. Part I: description and mathematical formulation." Preprint, Department of Petroleum Engineering, University of Texas, Austin.

Lanczos, C. (1938), "Trigonometric interpolation of empirical and analytic functions." *J. Math. Phys.* **17**, 123-199.

Lantz, R. B. (1970), "Rigorous calculation of miscible displacement using immiscible reservoir simulators." *Soc. Pet. Eng. J.* (June), 192-202.

Lantz, R. B. (1971), "Quantitative evaluation of numerical diffusion (truncation error)." *Soc. Pet. Eng. J.* (September), 315-320.

Lax, P. D. (1957), "Hyperbolic systems of conservation laws II." *Comm. Pure Appl. Math.* **10**, 537-566.

Lax, P. D. (1972), "The formation and decay of shock waves." *Am. Math. Monthly* (March), 227-241.

Li, Y.-K., and Nghiem, L. X. (1982), "The development of a general phase envelope construction algorithm for reservoir fluid studies." SPE 11198, presented at the 57th Annual Fall Technical Conference and Exhibition of the Society of Petroleum Engineers of AIME, New Orleans, September 26-29, 1982.

Lohrenz, J., Bray, B. G, and Clark, C. R. (1964), "Calculating viscosities of reservoir fluids from their compositions." *Jour. Pet. Tech.* (October), 1171-1176.

Mansoori, J. (1982), "Discussion of compositional modeling with an equation of state." *Soc. Pet. Eng. J.* (April), 202-203.

Mehra, R. K., Heidemann, R. A., and Aziz, K. (1980), "Computation of multiphase equilibrium for compositional simulation." SPE 9232, presented at the 55th Annual Fall Technical Conference and Exhibition of the Society of Petroleum Engineers of AIME, Dallas, September 21-24, 1980.

Mercer, J. W., and Faust, C. R. (1977), "The application of finite-element techniques to immiscible flow in porous media." In: *Finite Elements in Water Resources: Proceedings of the First International Conference on Finite Elements ins Water Resources* held in Princeton, New Jersey, July, 1976. London: Pentech Press, 1.21-1.58.

Metcalfe, R. S., Fussell, D. D., and Shelton, J. L. (1973), "A multicell equilibrium separation model for the study of multiple contact miscibility in rich-gas drives." *Soc. Pet. Eng. J.* (June), 147-155.

Metcalfe, R. S., and Yarborough, L. (1979), "The effect of phase equilibria on the CO_2 displacement mechanism." *Soc. Pet. Eng. J.* (August), 242-252.

Michelsen, M. L. (1980), "Calculation of phase envelopes and critical points for multicomponent mixtures." *Fluid Phase Equilibria* 4, 1-10.

Mohanty, K. K., and Salter, S. J. (1982), "Multiphase flow in porous media: II. Pore-level modeling." SPE 11018, presented at the 57th Annual Fall Technical Conference and Exhibition of the Society of Petroleum Engineers of AIME, New Orleans, September 26-29, 1982.

Morrow, N. R. (1969), "Physics and thermodynamics of capillary action in porous media." In: *Symposium on Flow Through Porous Media*, ed. by R. J. Nunge. Washington, D. C.: American Chemical Society, 104-128.

Mungan, N., and Johansen, R. T. (1974), "Miscible fluid displacement." In: *Secondary and Tertiary Oil Recovery Processes*, ed. by F. H. Poettmann, D. C. Bond, and C. R. Hocott. Tulsa: Interstate Oil Compact Commission.

Münster, A. (1970), *Classical Thermodynamics*, transl. by E. S. Halberstadt. New York: Interscience.

Nghiem, L. X., and Aziz, K. (1979), "A robust iterative method for flash calculations using the Redlich-Kwong or the Peng-Robinson equation of state." SPE 8285, presented at the 54th Annual Fall Technical Conference and Exhibition of the Society of Petroleum Engineers of AIME, Las Vegas, Nevada, September 23-26, 1979.

Nghiem, L. X., Fong, D. K., and Aziz, K. (1981), "Compositional modeling with an equation of state." *Soc. Pet. Eng. J.* (December), 687-698.

Nikolaevskii, V. N., and Somov, B. E. (1978), "Heterogeneous flows of multi-component mixtures in porous media -- review." *Int. J. Multiphase Flow* **4**, 203-217.

Nolen, J. S. (1973), "Numerical simulation of compositional phenomena in petroleum reservoirs." SPE 4274, presented at the Third Symposium on Numerical Simulation of Reservoir Performance of the Society of Petroleum Engineers of AIME, Houston, January 11-12, 1973.

Nunge, R. J., and Gill, W. N. (1969), "Mechanisms affecting dispersion and miscible displacement." *Ind. Eng. Chem.* **61**: 9, 33-49.

Oden, J. T., and Reddy, J. N. (1976), *An Introduction to the Mathematical Theory of Finite Elements*. New York: John Wiley and Sons.

Oellrich, L., Plocker, U., Prausnitz, J. M., and Knapp, H. (1981), "Equation-of-state methods for computing phase equilibria and enthalpies." *Int. Chem. Engrg.* **21**: 1, 1-16.

Olds, R. H., Reamer, H. H., Sage, B. H., and Lacey, W. N. (1949), "Phase equilibria in hydrocarbon systems. The n-butane - carbon dioxide system." *Ind. Eng. Chem.* **41**: 3, 475-482.

Oleinik, O. A. (1963a), "Construction of a generalized solution of the Cauchy problem for a quasi-linear equation of first order by the introduction of 'vanishing viscosity'." In: *American Mathematical Society Translations*, series 2, **33**. Providence: American Mathematical Society, 277-283.

Oleinik, O. A. (1963b), "Uniqueness and stability of the generalized solution of the Cauchy problem for a quasi-linear equation." In: *American Mathematical Society Translations*, series 2, **33**. Providence: American Mathematical Society, 285-290.

Orr, F. M., and Jensen, C. M. (1982), "Interpretation of pressure-composition diagrams for CO_2 - crude oil systems." SPE 11125, presented at the 57th Annual Fall Technical Conference and Exhibition of the Society of Petroleum Engineers of AIME, New Orleans, September 26-29, 1982.

Ortega, J. C., and Rheinboldt, W. C. (1970), *Iterative Solution of Non-Linear Equations in Several Variables*. New York: Academic Press.

Page, R. H. (1982), "A Fortran utility program for triangular finite elements: grid checking, grid plotting, and solution contour plotting." Unpublished computer code documentation, Department of Civil Engineering, Princeton University, Princeton, New Jersey.

Peaceman, D. W. (1977), *Fundamentals of Numerical Reservoir Simulation*. Amsterdam: Elsevier Scientific Pub. Co.

Peng, D.-Y., and Robinson, D. B. (1976), "A new two-constant equation of state." *Ind. Eng. Chem. Fundam.* **15**: 1, 53-64.

Peng, D.-Y., and Robinson, D. B. (1977), "A rigorous method for predicting the critical properties of multicomponent systems from an equation of state." *AIChE Jour.* **26**: 2, 137-144.

Peng, D.-Y., and Robinson, D. B. (1980), "Discussion of critical point and saturation pressure calculations for multicomponent systems." *Soc. Pet. Eng. J.* (April), 75-80.

Perkins, T. K., and Johnson, O. C. (1963), "A review of diffusion and dispersion in porous media." *Soc. Pet. Eng. J.* (March), 70-84.

Pinder, G. F., Frind, E. O., and Celia, M. A. (1978), "Groundwater flow simulation using collocation finite elements." In *Proceedings of the Second International Conference on Finite Elements in Water Resources*, London, July, 1978. London: Pentech Press, 1.171-1.185.

Pinder, G. F., and Gray, W. G. (1977), *Finite Element Simulation in Surface and Subsurface Hydrology*. New York: Academic Press.

Poling, B. E., Grens, E. A., and Prausnitz, J. M. (1981), "Thermodynamic properties from a cubic equation of state: avoiding trivial roots and spurious derivatives." *Ind. Eng. Chem. Process Des. Dev.* **20**, 127-130.

Pope, G. A., Carey, G. F., Lake, L. W., and Sepehrnoori, K. (1982), "Isothermal, multicomponent multiphase fluid-flow in permeable media. Part II: numerical techniques and solution." Preprint, Department of Petroleum Engineering, University of Texas, Austin.

Prenter, P. M. (1975), *Splines and Variational Methods.* New York: John Wiley and Sons.

Prevost, J. H. (1980), "Mechanics of continuous porous media." *Int J. Engrg. Sci.* **18**, 787-800.

Price, H. S., and Donohue, D. A. T. (1967), "Isothermal displacement processes with interphase mass transfer." *Soc. Pet. Eng. J.* (June), 205-215.

Raimondi, P., and Torcaso, M. A. (1965), "Mass transfer between phases in a porous medium: a study in equilibrium." *Soc. Pet. Eng. J.* (March), 51-59.

Reamer, H. H., and Sage, B. H. (1963), "Phase equilibria in hydrocarbon systems. Volumetric and phase behavior of the n-decane - CO_2 system." *J. Chem. Eng. Data* **8**: 4, 508-513.

Redlich, O., and Kwong, J. N. S. (1949), "On the thermodynamics of solutions. V. An equation of state. Fugacities of gaseous solutions." *Chem. Rev.* **44**, 233-244.

Reid, R. C., and Beegle, B. L. (1977), "Critical point criteria in Legendre transform notation." *AIChE Jour.* **23**: 5, 726-732.

Reid, R., Prausnitz, J., and Sherwood, T. (1977), *The Properties of Gases and Liquids,*" 3rd ed. New York: McGraw-Hill.

Risnes, R., Dalen, V., and Jensen, J. I. (1981), "Phase equilibrium calculations in the near-critical region." In: *Proceedings of the Third European Symposium on Enhanced Oil Recovery*, Bournemouth, U.K., ed. by F. J. Fayers. Amsterdam: Elsevier, 329-350.

Roebuck, I. F., Henderson, G. E., Douglas, J., and Ford, W. T. (1969), "The compositional reservoir simulator: case I -- the linear model." *Soc. Pet. Eng. J.* (March), 115-130.

Russell, R. D., and Shampine, L. F. (1972), "A collocation method for boundary value problems." *Numer. Math.* **19**, 1-28.

Sandrea, R., and Nielsen, R. F. (1974), *Dynamics of Petroleum Reservoirs Under Gas Injection*. Houston: Gulf Pub. Co.

Scheidegger, A. (1974), *The Physics of Flow Through Porous Media*, 3rd ed. Toronto: University of Toronto Press.

Settari, A., Price, H. S., and Dupont, T. (1977), "Development and application of variational methods for simulation of miscible displacement in porous media." *Soc. Pet. Eng. J.* (June), 228-246.

Shapiro, A. M., and Pinder, G. F. (1981), "Analysis of an upstream weighted collocation approximation to the transport equation." *Jour. Comput. Phys.* **39**, 46-71.

Shapiro, A. M., and Pinder, G. F. (1982), "Solution of immiscible displacement in porous media using the collocation finite element method." In *Proceedings of the Fourth International Conference on Finite Elements in Water Resources*, Hannover, F. R. G., June 1982. Berlin: Springer-Verlag, 9.61-9.70.

Sigmund, P. M., Dranchuk, P. M., Morrow, N. R., and Purvis, R. A. (1973), "Retrograde condensation in porous media." *Soc. Pet. Eng. J.* (April), 93-104.

Sincovec, R. F. (1977), "Generalized collocation methods for time-dependent, nonlinear boundary-value problems." *Soc. Pet. Eng. J.* (October), 345-352.

Smith, B. J. (1967), "ZERPOL, a zero finding algorithm for polynomials using Laguerre's method." Department of Computer Science, University of Toronto, Toronto, Ontario.

Smith, B. L., and Schwartz, F. W. (1980), "Mass transport I. A stochastic analysis of macroscopic dispersion." *Water Res. Research* **16**, 303-313.

Smith, C. R. (1966), *Mechanics of Secondary Oil Recovery*. Huntington, N.Y.: Robert E. Krieger Pub. Co.

Spivak, A. Price, H. S., and Settari, A. (1977), "Solution of the equations for multidimensional, two-phase, immiscible flow by variational methods." *Soc. Pet. Eng. J.* (February), 27-41.

Stalkup, F. I. (1978), "Carbon dioxide miscible flooding: past, present, and outlook for the future." *Jour. Pet. Tech.* (August), 1102-1112.

Strang, G., and Fix, G. J. (1973), *An Analysis of the Finite Element Method*. Englewood Cliffs, N. J.: Prentice Hall.

Taylor, G. I. (1953), "Dispersion of soluble matter in solvent flowing slowly through a tube." *Proc. Roy. Soc. Lond. A* **219**, 186-203.

Temple, J. B. (1981), "Global existence for a class of 2x2 nonlinear conservation laws with arbitrary Cauchy data." Preprint, Department of Mathematical Physics, Rockefeller University, New York.

Todd, M. R., and Longstaff, W. J. (1972), "The development, testing, and application of a numerical simulator for predicting miscible flood performance." *Jour. Pet. Tech.* (July), 874-882.

Van Genuchten, M. T., and Gray, W. G. (1978), "Analysis of some dispersion corrected numerical schemes for solution of the transport equation." *Int. J. Num. Meth. Engrg.* **12**, 387-404.

Van Quy, N., Simandoux, P. and Corteville, J. (1972), "A numerical study of diphasic multicomponent flow." *Soc. Pet. Eng. J.* (April), 171-184.

Varotsis, N., Todd, A. C., and Stewart, G. (1981), "An iterative method for phase-equilibria calculations with particular applications to multicomponent miscible systems." In: *Proceedings of the Third European Symposium on Enhanced Oil Recovery*, Bournemouth, U. K., ed. by F. J. Fayers. Amsterdam: Elsevier, 313-328.

Watkins, R. W. (1978), "A technique for the laboratory measurement of carbon dioxide unit displacement efficiency in reservoir rock." SPE 7474, presented at the 53rd Annual Fall Technical Conference and Exhibition of the Society of Petroleum Engineers of AIME, Houston, October 1-3, 1978.

Watkins, R. W. (1982), "The development and testing of a sequential, semi-implicit four component reservoir simulator." SPE 10513, presented at the Sixth SPE Symposium on Reservoir Simulation held in New Orleans, February 1-3, 1982.

Welge, H. J. (1952), "A simplified method for computing oil recovery by gas or water drive." *Trans. AIME* **195**, 97-98.

Wolfe, P. (1959), "The secant method for simultaneous nonlinear
equations." *Commun. ACM* **2**: 12, 12-13.

Yellig, W. F., and Baker, L. E. (1981), "Factors affecting miscible
flooding dispersion coefficients." *Can. J. Pet. Tech.* (October-
December), 69-75.

Young, L. C., and Stephenson, R. E. (1982), "A generalized compositional
approach for reservoir simulation." SPE 10516, presented at the Sixth
SPE Symposium on Reservoir Simulation, New Orleans, February 1 - 3,
1982.

Lecture Notes in Engineering

Edited by C.A. Brebbia and S.A. Orszag